Ancestors for the Pigs: Pigs in Prehistory

MASCA Research Papers in Science and Archaeology

Series Editor
Kathleen Ryan

MASCA Research Papers in Science and Archaeology

Volume 15, 1998

Ancestors for the Pigs: Pigs in Prehistory

edited by
Sarah M. Nelson

Museum Applied Science Center for Archaeology
University of Pennsylvania Museum of Archaeology and Anthropology
1998

Published by
Museum Applied Science Center for Archaeology (MASCA)
University of Pennsylvania Museum of Archaeology and Anthropology
33rd and Spruce Streets, Philadelphia, PA 19104-6324
www.upenn.edu/museum

Copyright 1998

ISSN 1048-5325

Printed by Cushing-Malloy, Inc.
Ann Arbor, Michigan

Cover illustration:
Hongshan culture jade pig-dragon. Image courtesy of
Guo Da-shun, Director of the Liaoning Province Archaeology
and Cultural Relics Institute, Shenyang, China. Redrawn by
Ardeth Abrams from *The Archaeology of Northeast China*
(London: Routledge 1995), 37.

TABLE OF CONTENTS

Preface.. vii
Sarah M. Nelson

Introduction 1
Sarah M. Nelson

Reflections on the Interactions Between People and Pigs 5
Alice Dawson

Privy-Pigs in Prehistory? A Korean Analog for Neolithic Chinese Subsistence Practices 11
David J. Nemeth

An Ethnographic View of the Pig in Selected Traditional Southeast Asian Societies 27
P. Bion Griffin

Evaluation of Molar Size as a Basis for Distinguishing Wild Boar from
Domestic Swine: Employing the Present to Decipher the Past 39
John J. Mayer, James M. Novak, and I. Lehr Brisbin, Jr.

Early Pig Husbandry in Southwestern Asia and Its Implications for
Modeling the Origins of Food Production 55
Michael Rosenberg and Richard W. Redding

Ancestral Pigs: A New (Guinea) Model for Pig Domestication in the Middle East 65
Richard W. Redding and Michael Rosenberg

Pig Exploitation at Neolithic Çayönü Tepesi (Southeastern Anatolia) 77
Hitomi Hongo and Richard H. Meadow

Pigs in the Hongshan Culture 99
Sarah M. Nelson

Pigs and Emergent Complexity in the Ancient Near East... 109
Melinda A. Zeder

Pig Use and Abuse in the Ancient Levant: Ethnoreligious Boundary-Building with Swine 123
Brian Hesse and Paula Wapnish

Pigs in Ancient Egypt 137
Richard A. Lobban

PREFACE

Why a book about pigs in prehistory? This book began as an invited symposium for the AAA in San Francisco in 1996. The scholarly reasons for organizing the symposium can be found in my introductory chapter. Here I would like to be more lighthearted, with a few comments about why we archaeologists have had limited models for understanding pigs in the distant past.

In current American culture, the proper destiny of pigs is to appear on the table, as in the folk song in which the mother pig sings, "All my sons and all my daughters are hams and hocks and bacon and trotters." This highlights the fact that in America, pigs are simply meat—bacon, pork, and ham (Babe, Wilbur, Porky, and Piglet notwithstanding). Few of us have met a pig on the hoof (or on the trotter, more appropriately); they are most frequently encountered as parts wrapped in see-through packaging in the supermarket cold case. In spite of the fact that few of us ever encounter live pigs, notions about them as farm animals pervade the culture. Pigs are a metaphor for dirt and dung, mess and mud.

European traditional cultures kept pigs and ate them with gusto (and still do), but pigs do not often feature in folklore, still less do they have anything to do with ideology. When they are found in folktales, as in "The Three Little Pigs," they are still food—for the wolf. Thus, we are equally bemused by both the "pig lovers and pig haters," as Marvin Harris (in *Cows, Pigs, Wars and Witches* [New York: Vintage Books, 1974]) has dubbed the cultures that feast on pigs and those that prohibit the consumption of pork.

However, pigs are more than food even in a utilitarian sense. Not only are inedible pig parts (skin, bones, bristles, etc.) useful for making various products, but pigs can be important in waste management. Pigs will consume both human excreta and kitchen refuse that human digestive tracts cannot handle at all. Perhaps this makes them seem "dirty," but in fact they can be a help in keeping a village clean and in suppressing disease.

Pigs have different uses in different places, and those uses are related to both the local ecology and the local ideology. Pigs in Asia, like those elsewhere, are a source of food. Chinese cuisine in particular would not be complete without pork. Chinese cookbooks offer dozens of ways to prepare pork—wrapped in thin pancakes, stuffed in triangular pastry, flavored sweet and sour, and on and on. Any aficionado of Chinese restaurants, any champion of chopsticks knows many of the varieties. Pigs also are found in Japanese cuisine, but much less commonly in that culture, where products of the sea are the most esteemed sources of protein. Traditional Korean food also included pork, often in soups, but pork is eaten less there than in China, although more than in Japan.

However, in Asia the family pig is perhaps at least as important as a waste disposal unit. Pigs are omnivorous, and will eat many things that humans will not or cannot eat. The coupling of the pigsty with the latrine is common in both China and Korea. Models of pigsties with latrines found in noble tombs of the Han Dynasty in China (second century B.C. to second century A.D.) show that they were important enough to take along into eternity. In Korea, the humor of many scatological stories revolves around this association. If the latrine door is closed, one does not simply holler out to inquire if the facility is occupied. That would be extremely impolite. The proper announcement is a discreet cough, and if someone is there, they cough back. Of course, a pig's oink can sound like a cough, leaving the family member in a quandary, desperately coughing outside the door and hearing answering grunts.

Where pigs are not penned up to recycle human waste, they may be allowed to forage for food, especially where there is forest, the natural habitat of wild pigs. They may also be turned loose in the streets of villages, as they are in Jain villages in Rajasthan, India. Just like "sacred cows," pigs wander among the traffic, finding tasty bits of offal. They tend to range freely in the Chinese countryside as well, especially in locations far from cities. Sometimes a person is present to herd them, but often they are on their own, untended, expected to find their own way home.

Pigs in Asia are usually kept by the woman of the house and are considered her property. It is work to feed a pig, especially as it grows larger. Any increase (in the form of growth or piglets) is recognized as interest on the labor she has invested. Sale of the grown pig or the piglets may enable her to obtain some special luxury, such as the woman documented in the film "Small Happiness," who raised piglets to purchase a tape player, and listened to Chinese opera happily thereafter. Thus pigs are connected with luxury consumption, either directly or indirectly—either pork feasts or the extras they provide in a money economy. It is no accident that the "piggy bank" is a product of China, for pigs there are a symbol of wealth.

Asian pigs are taken to market, but they are sold in nearby market towns, not as part of long-distance trade. When pigs are fattened up for sale they are walked only when other transportation is not available. Why walk off all those valuable calories? I have seen pigs transported to market in three-wheeled trucks, in contraptions pulled by tractors, and even draped over a bicycle. Pigs may be carried on the back in some cases.

These brief examples are intended to demonstrate that we have much to learn about the pigs whose bones or images are found in archaeological sites, and particularly that we should consider more than the caloric value of pigs. Our own culture places blinders on us in our approaches to ancient pigs.

I would like to thank all the speakers in the AAA symposium, especially those who revised their papers for this book, as well as the additional chapters from those who could not participate in the symposium. Many thanks as well to Kathleen Ryan, Series Editor of the MASCA Research Papers, who heroically tried to herd each of us into producing our chapters in this book. I hope it will be useful to everyone who identifies pigs in their sites.

Sarah M. Nelson
September 1998

INTRODUCTION: PIGS IN PREHISTORY

Sarah M. Nelson

Dept. of Anthropology, 2130 S. Race, University of Denver, Denver, CO 80208

Archaeology has adopted many of its models for understanding the past from ethnographies. Ethnographic analogy can be a powerful tool for archaeological interpretation. Thus, I have inverted a well-known ethnographic title, *Pigs for the Ancestors* by Roy Rappaport, for this volume about pigs in the prehistoric past, because the rearranged title speaks to that point. *Pigs for the Ancestors* is familiar to most anthropologically trained archaeologists; almost too familiar. Its images—of pig feasts, of pigs eating the same food as humans and becoming increasingly difficult as they become too big and too many for the women who tend them, of pigs as wealth and in competitive feasting for allies in warfare—have all struck the archaeological imagination and provided explanations for far-away cultures wherever there are pigs.

At times it has seemed to me that the notion of "feasting" has been stretched by archaeologists to explain too much (Nelson 1997:141–145). In particular, the role of pigs in that process is too limited a model to cover the very multitudinous and diverse ways that pigs have interacted with humans (Nelson 1994; Nemeth 1995). In spite of the frequency of evidence of pigs in archaeological sites throughout Eurasia from Neolithic times, models for interpreting pigs in prehistory are extremely limited. Worse yet, these models have often been insensitive to cultural and ecological differences between regions and sites. This book is intended to broaden the archaeologists' awareness and understanding of pigs, not just as providing food but in serving other functions as well. Pigs tend to be dismissed from the "Secondary Products Revolution" (Sherratt 1983) for they provide neither traction nor milk nor wool, but they have other non-meat uses. Many factors need to be considered in addition to the fact of the presence of pigs. Archaeological interpretations of the presence of pig bones and pig images in sites can be improved by making them more nuanced and more context-specific. Understanding the range of pig/human interactions in present societies is a partial solution. In addition, specific ethnographic analogies can be selected and applied with more precision.

The title, "Ancestors for the Pigs," thus refers not only to evidence for early pig/human interaction, but also to the "ancestors" of the models archaeologists use. This book brings together several new ways of thinking about pigs in the past, creating a dialogue by drawing on several kinds of approaches—from geography, ethnography, zoology, history, and archaeology—to enrich the way we all understand the evidence found in archaeological sites.

Other models besides New Guinea (viz. Rappaport) may be useful, and additional detailed information that sheds light on evidence of pigs—images of pigs as well as pig bones—in various archaeological contexts—burials, pits, houses; rural and urban—needs to be weighed in preferring any one interpretation over another. Furthermore, thinking about the needs of pigs might lead to better interpretations of structures in archaeological sites. Starting with the pig itself may well produce insights that might otherwise be overlooked. All of these approaches have been taken by various authors in this book.

Many questions should be considered when interpreting evidence of pigs in archaeological sites. When were pigs first domesticated, and how did they fit into the lifeways of the domesticators? What did pigs eat in climates as different as northern China, southern China, and Anatolia? Did pigs need to be tended? If so, who herded them? If not, how were they kept from straying? Where did pigs sleep? How closely did pigs live with humans? Were they sacrificed? What uses did they have in addition to being human food?

Other questions are raised by these chapters. What might the ecology of pigs tell us cross-culturally? What specific uses of pigs might be evident in a particular site? Why were pigs an early domesticate throughout their range, and how did they become preferred or despised? What did pigs mean in the cultures whose remains we excavate? What are the implications for an animal such as the pig that can't be herded the way that sheep, goats, and cattle can? What are the implications for human organization? Were there categories of people designated to tend the pigs? Swineherds? Women? Children? These questions begin to be addressed in the following chapters.

Organization of the Book

The volume begins with America in the present, and a geographic approach to pig/human interaction. This perspective is extremely valuable for what it reveals about our own biases, as they might be applied unwittingly and inappropriately to pigs in the past. Alice Dawson demonstrates that as the population has become less and less rural, the American approach to pigs has been more and more industrial. This has meant not merely treating pigs only as meat, but mass producing them in ways that make it clear that efficiency (for humans) and therefore profit are the chief considerations. The storied pigs—Piglet and Babe, for example—have no place in these pig-raising factories, even though cute pigs may be used in advertisements.

While Dawson lays out the realities of pigs in the present, Jim Nemeth points out the consequence for archaeological interpretations of the American mindset that considers pigs only as food. He shows that on Cheju Island in Korea, even though pigs may eventually be eaten, they functioned in the recent past as transformers of human waste into fertilizers as well as meat, with the pigsty-privy an important part of the ecosystem. Furthermore, in considering the role of pigsty-privies in the ecosystem of disease, rather than being an unsanitary practice that promoted illness, they helped keep some endemic parasites in check. This example of the need to extend our notions about pigs is especially useful as a cautionary tale for archaeological explanations involving the role of pigs in early domestication, as well as their place in urban economies and the rise of complex societies.

Bion Griffin provides an object lesson in noting the variety of ways that pigs may be hunted or domesticated, depending on a number of variables. He uses detailed examples of peoples and pigs from the ethnographic record in Southeast Asia. The preference of some groups for wild boar over domestic pigs is especially important for conceptualizing the beginnings of pig domestication. If wild pigs are plentiful and easily hunted, why domesticate them? Griffin's work also sheds light on the archaeological sites that contain both wild and domestic forms of pig. Furthermore, he shows that the wild bearded pigs of Southeast Asia actually migrate in large herds, raising questions about the possible migration of other varieties of pigs. Perhaps pigs could be herded after all.

How to distinguish wild boar from domesticated pigs is a continuing problem for archaeologists. Understanding the meanings of pig bones in a site often depends upon whether they represent wild boar or domesticates. In several sites the presence of unusual numbers of young pigs has been argued to indicate the culling of young males, and therefore domestication. John Mayer, James Nova, and Lehr Brisbane demonstrate that second and third molars are particularly useful for distinguishing wild from domestic pigs, since teeth are often preserved even when other faunal remains are not. The molar teeth may also elucidate the process of domestication, as the changes are gradual.

The remainder of the chapters deal with archaeological sites and regions. The chapters are arranged according to stages of complexity rather than regions. Instead of moving around the world from China to Egypt (or the other way around), the chapters progress from early evidence of domestication to pigs in complex societies. Thus the only chapter on the archaeology of East Asia appears in the midst of the chapters on southwest Asia and North Africa. The earlier three chapters look at the beginnings of pig domestication, while the later three address questions of pig avoidance and consider closely the meanings of the remains found in archaeological sites.

Two chapters take as their main theme the domestication of swine. Mike Rosenberg and Richard Redding discuss the site of Hallan Çemi in eastern Turkey, with its evidence for early pig domestication. They suggest that in fact pigs are the earliest animal domesticate in Anatolia. This substantially revises the understanding of animal domestication in this region. The discovery of domesticated pigs in an aceramic site parallels the earliest dated pigs in China. The southern Chinese site of Zengbiyan is also small but sedentary, with burials and architecture in a cave (Chang 1986:101–105). Careful comparison of these sites might be productive.

In their second chapter, Redding and Rosenberg consider the application of New Guinea ethnography, including *Pigs for the Ancestors*, as a model for explaining the data in Anatolia. They consider a model that could be thought of as domestication of female pigs, but the continuing wildness of boars. This model resonates with the ethnographic examples in Griffin's chapter. Furthermore, there was a source of food for pigs (acorns) that the human population did not use. It is germane that pigs were not the most common animal found at the site, although they were domesticated. Both deer and sheep/goats are more common. However, more pigs were butchered at the site than any other animal. Presumably hunted animals tended to be represented by meat bones rather than whole skeletons, suggesting that they were killed and butchered elsewhere. Competitive feasting is proposed based on finer and more elaborate objects found in the upper layers. These authors also suggest that when people took up cereal cultivation seriously, pigs would have been a liability, as they are prone to trampling crops.

A later step in pig domestication is discussed by Hitomi Hongo and Richard Meadow, using zoological data from the site of Çayonu. Some relevant data from Japan are brought into the discussion as well. Again, a mix of wild and domesticated pigs seems to be indicated. This finding is not often considered in models of domestication, but these chapters show that it is often important. Chapters that focus on complex societies consider both ideology and ecology.

The China chapter considers pig iconography as it developed along with complexity. This was not at the beginning of pig domestication, but occurred after both pig bones as well as figurines of pigs had been found in northern China for roughly 2000 years. The association of pigs and ideology is given an ecological slant, but it is nevertheless striking that pigs are important in the iconography of the Hongshan culture, with pig heads on what may be the first "dragons" in China.

On the other edge of Eurasia, where pig taboos became strong, two chapters look at the archaeological evidence for the avoidance of pigs, and both conclude that ethnicity is not the sole answer. Mindy Zeder considers the variable of urbanism in the utility of pigs for various populations. It is particularly telling that a kibbutz near one of the sites Zeder discusses actually raises pigs. Brian Hesse and Paula Wapnish use biblical texts to consider the taboos and the ways that taboos have been variously observed and ignored. They show that even pig avoidance is not simple, but has many permutations.

Moving to North Africa, the use of pigs in Egypt is discussed by Richard Lobban. His data are particularly intriguing, because Egypt is not generally thought of in connection with pigs. Yet he shows that pigs were raised along the Nile throughout prehistory. Lobban raises the important question of pig feed, and the raising of cereals specifically for fodder. When pigs could no longer range freely to find their own food, because the marshes that supported them were being transformed into agricultural land, they became less useful to humans, a fact that is reflected in the local mythologies.

Themes and Contrasts in the Chapters

Several themes that can be extracted from these chapters are worth considering further. These include the functions of pigs, the characteristics of pig keepers, spatial aspects of pigs, food for pigs and pigs as food, and various facets of pigs in ideology. Some comparisons between the chapters are particularly revealing of new patterns of thinking about pigs in prehistory.

The Functions of Pigs

Pigs are not just meat, even in contemporary society. Alice Dawson points out that designer pigs can even be pets and companions. She also details the medical uses of pigs, which are often little known; much research is being undertaken with pigs for the benefit of humans. In China, pig parts such as tusks and internal organs are believed to have medicinal value in themselves (Zhang 1996). Jim Nemeth emphasizes pigs as waste collectors and fertilizer producers, and as useful for disease control. Richard Lobban mentions that pigs may be used deliberately to trample seeds into the soil. These observations broaden the ways pigs can be thought about in early sites.

Gender, Class, and Ethnicity

My chapter notes that in China pig keepers are often women, and Jim Nemeth confirms that this is also true in Korea. Although no gender connection is made in any other paper, this finding coincides with the data from New Guinea, where the pig raisers are also women (Rappaport 1967). Ethnicity, on the other hand, is considered in several chapters. Hesse and Wapnish show that Jewish ethnicity defined in terms of pork prohibition varied through time. Groups surrounding the Jews and their own interpretations of ethnicity also played a part. Using a combination of site data and biblical texts, Hesse and Wapnish demonstrate that the avoidance of pigs changed through time for the Israelites. Lobban also makes distinctions between various ethnic groups in Egypt and their observance, or non-observance, of the pig taboo. Zeder notes that class differences are relevant in considering who raises swine. Thus, the characteristics of pig keepers are useful data for interpreting pigs of the past.

Pigs are kept in various places when they are domesticated. Dawson describes the appalling conditions of pigs in commercial feed lots, but these do not apply to prehistory. Nemeth discusses the keeping of pigs in pens under privies, and Griffin describes pigs living under houses in the Philippines. Both emphasize the role of pigs as foragers. In China, the character for house/family consists of a pig under a roof, suggesting the ubiquity of the pig as a dweller in the same space as people.

Food for Pigs and Pigs as Food

Griffin relates that pigs are fed gathered foods in the Philippines, in addition to being allowed to forage for garbage within the settlement. Richard Lobban shows that in Egypt pigs may even be fed cultivated plants, such as emmer wheat. Otherwise, little is said about what pigs eat, except that the remains of human food, as well as human waste, are often designated as pig food. Pigs as meat are considered tasty wherever consumption is permitted. Griffin suggests that people need the lard from pigs, which may be especially true where other fat sources are scarce or lacking.

Ecology of Pigs

Pigs tend be domesticated when people live in settlements. The conclusion therefore seems to be that pigs are not suited to nomadic life. Zeder tests several ecological theories of pig presence or absence in southwest Asian sites. She looks at aridity, pastoral production, and agricultural intensification at a single site to test which is definitive. She finds that when nomads are doing much of the meat provisioning, pigs are poorly represented, but the controlling factor is the immersion of the site in a larger economy. A measure of economic autonomy in urban situations may be the deciding factor, especially for people in marginal economic circumstances. In my chapter I suggest that the increasing desiccation that occurred during the time of the Hongshan culture may have led to an increased investment in pigs as symbols. The difference in ecology between Upper Egypt and the Nile Delta region is used by Lobban to explain differing attitudes toward pigs.

Pig Iconography

Other than the chapter that focuses on images of pigs (Nelson), there are few examples of pig iconography. However, Lobban notes that pigs on cylinder seals, as well as faience pigs, are found in Egypt. The pig in Egypt represented the god Seth in the Lower Nile Delta, but this was often not a positive image. It seems that in southwest Asia and North Africa the pig was seldom celebrated with images, nor was it used for burial offerings as it was in Asia. In the Hongshan culture in China, pig representations become less realistic over time, and imply transcendent meanings.

Pigs and Mythology

Pigs have exalted places within some mythologies, and mixed characteristics in others. In China, a pig is a major character in the beloved story, "Journey to the West," which also features a character called Monkey. Pig is one of the immortals, although he is portrayed as a glutton and a coward (Zhang 1996). The nomarchs of the Lower Delta used the pig as a totem, but their conquest by Upper Egypt made pigs low-status animals. While the southwest Asia chapters tend to consider the low status of pigs, this was not always the case. For example, Lobban reveals that pigs were used to propitiate the supernatural at the Temple of Osiris in Egypt. Alice Dawson shows us that pigs are not necessarily considered to be lowly in Europe, either. In Scandinavia, a "great divine pig" pulled the chariot of the god Frey and was one of the immortals in Valhalla. We may find it amusing that pigs pull Santa's sleigh in Sweden, but it is useful to keep in mind the multiple attitudes toward pigs and dimensions of pigs, as the following chapters are digested.

References

Chang, K.C. 1986. *The Archaeology of Ancient China.* Yale University Press, New Haven.

Nelson, Sarah. 1994. Comments on "Burials, Pigs, and Political Prestige in Neolithic China." *Current Anthropology* 35(2): 135–136.

——— 1997. *Gender in Archaeology: Analyzing Power and Prestige.* AltaMira Press, Walnut Creek, CA.

Nemeth, David J. 1995. On Pigs in Subsistence Agriculture. *Current Anthropology* 36(2): 292–293.

Rappaport, Roy. 1967. *Pigs for the Ancestors: Ritual in the Ecology of a New Guinea People.* Yale University Press, New Haven.

Sherratt, Andrew. 1983. The Secondary Exploitation of Animals in the Old World. *World Archaeology* 15(1): 90–104.

Zhang, Ka Bo. 1996. Chinese Pig Tales. *Archaeology* 49(2): 52–57.

REFLECTIONS ON THE INTERACTIONS BETWEEN PEOPLE AND PIGS

Alice Dawson

Department of Geography, University of North Carolina at Chapel Hill, Saunders Hall, Chapel Hill, NC 27599-3220

Animals and Geography

Animals, human and non-human, are embedded in the natural biophysical environment. The non-human animal species are an essential element of human material culture. They are a source of food, clothing, personal care products, medical products, physical assistance for the disabled, friendship and companionship. Animals are an important part of our vocabulary and symbolism. Humans develop relationships with other animals that are unique, that do not occur with any other component of the human environment. In short, animals are an element of the biophysical environment that is modified in numerous ways by human activity, particularly as animals are incorporated into the cultural environment.

Geography is concerned with cultural-environmental interaction—the ways in which humans affect and are affected by the biophysical environment—and spatial pattern and variation in the environment, and human use of this environment. This includes human perceptions of and meanings given to environments and the consequent use and modification of that environment, and the use of space. With this approach to organizing human knowledge and understanding of the world, geography has much to offer in considerations of the human/animal relationship (Bennett 1960; Simoons 1974; Vale and Parker 1980; Veblen 1989). Many unexplored opportunities to examine animals as the intersection of cultural and environmental interaction await.

This essay considers aspects of the interactions between human culture and one particular species in the biophysical environment: the pig. The primary focus is the domestic pig, examined within the Western cultural context. Additionally, the essay suggests a means to explore the human/animal relationship within geographical study by examining animals as a point of intersection between human culture and the biophysical environment. Consideration is given to the consequences of this interaction on the environment, particularly the aspect most affected—the animal itself. The discussion will also examine how humans have changed this element and restructured the space shared by humans and pigs as a consequence of their perception of the pig.

Pigs in Particular

Since at least the Middle Ages, pigs within Western culture have had and continue to have multiple roles: food, agricultural commodity, medical resource, working animal, and even companion. They are laden with symbolic meanings. Kearney (1981) and Sillar and Meyler (1961) examine many of these meanings and symbolisms throughout Western tradition. Pigs are very much a part of our vocabulary. Hedgepeth writes: "There are, in fact, more hog-related expressions than expressions involving any other animal—at least in the English language—and most of them have to do with linking men to hogs" (1978:39). Pigs are incorporated into various religious traditions, from the strong prohibitions of pork consumption within the Jewish and Muslim faiths, to their varied representations in Christian religious art (Sillar and Meyler 1961).

The meaning of pigs to humans and the ways in which we use the animal vary considerably through space and time. This essay considers the Western cultural tradition, primarily late twentieth century America, and the ways in which Americans interact with domestic pigs. How do we as Americans at the end of the 1990s perceive this part of our environment? How do we respond to and modify it as a consequence of our perceptions? What effect does this have on the spatial ordering of our biophysical environment?

The human-pig relationship is marked by considerable ambiguity, resulting in a spectrum of interactions with this animal in late twentieth century America. The consequences of this ambiguity have an effect on the spatial ordering and patterning of human activities as well. Three particular examples of ways that humans perceive and consequently use and modify pigs, and our structuring of common space in response to these perceptions, follow.

It is important to note that human perceptions of pigs quite likely also indicate something of our understanding of nature more broadly. This is a complex issue, beyond the scope of this essay, yet some aspects of this understanding are a part of the context in which humans interact with pigs. Western science and scientific thought play a role in our perception of pigs (as well as other animals) as a part of nature. In the post-medieval period, the dichotomy between reason and emotion crystallized, and the universality of science and generalization began to dominate particularities and individualism. Relph (1981) sees the anthropocentrism of scientific humanism as promoting the domination and exploitation of the natural environment. This plays a major role in the more recent modifications of our landscapes, both the natural and human-built.

The French philosopher Descartes should be mentioned in connection with the place of animals within the Western scientific framework. For Descartes animals lack the supreme faculty of reason and therefore are unfeeling and mechanical; they are machines (Descartes, in Regan and Singer 1976). The cries of a vivisected animal being probed are the same as the sounds made by the internal mechanisms of any other machine when touched. If nature and particularly animals are simply machines, without reason or soul, human behavior towards animals requires no ethical concern. They may be dominated, controlled, changed, and otherwise used with no regard for any perspective other than that of humans.

Tuan (1984) writes of humans creating not only gardens in their image of what nature should be, but also animals (including other humans). The animals are made to reflect the human understanding of what nature should be, often with unfortunate consequences for the species that are modified. He describes goldfish bred with huge bulging eyes, considered unusual and beautiful to some human goldfish fanciers. Unfortunately, these goldfish have not adapted to this change and injure their eyes by bumping into objects, which often leads to blindness. Since animals are considered to be beyond the sphere of human consideration because they have no reason, humans need have no hesitation in changing animals as desired.

Pork: Replacement Parts, or Pet?

Although pigs are living beings in the environment, we have made them part of our culture. We continue to try to shape and reshape them in the image we desire, as we incorporate them into human culture.

In his consideration of the American landscape, David Lowenthal writes: "The American scene, as much as any other, mirrors a long succession of idealized images and visual stereotypes" (1968:62), and he quotes George Perkins Marsh that while we "think that the earth made man, man in fact made the earth" (Lowenthal 1968:61). Humans modify the landscape and environment as an entity, but do so by reshaping individual elements to meet a desired image. Like other components of the biophysical environment, animals are modified based on human perception, need, and whim.

Through time, pigs have been viewed in different ways within the Western cultural tradition. Kearney (1991) examines the evolution of these various threads of thought through the Middle Ages. In Scandinavian legend, the great divine pig Gullinbursti (translated "Golden Bristle") pulled the chariot of Frey, the fertility god and patron of Sweden and Iceland, and mingled with other immortals at Valhalla. Remnants of this tradition linger in the boar's head celebration and carols that are part of the English Christmas, and pigs still pull Santa's sleigh in Sweden. Yet pigs were burned at the stake as witches at the height of the European witch hunts, and tried in courts of law for crimes (Nissenson and Jonas 1992). The wild boar was seen as brave and courageous, a powerful and cunning warrior, and a most worthy adversary. The domestic pig became an image of sloth, gluttony, and lasciviousness. The boar recurs as a symbol of nobility, the domestic pig as embodying the attributes of the slovenly peasant (Kearney 1991). To quote Kearney:

> This trip back to the Middle Ages of the pig has not only called attention to an important but obscured segment of medieval symbolism, but has also brought to the surface questions of sensuality and idealism, religious belief, views of class divisions, and the nature of humans. It has pointed out that the object of man's peculiar cultural disdain for the pig is less the beast itself than man's own speckled soul. (1991:322)

The pig carries a heavy burden indeed.

Americans in the 1990s continue in this mixed response to pigs. The ambiguity of our relationship with pigs is manifested in varied ways. One useful illustration is the marketing of pork products as food. The flesh of pigs as food is frequently marketed with smiling, dancing piglets. For example, the barbecue restaurant advertisements in any North Carolina community's yellow pages inevitably turn up a variety of "cute," usually smiling pigs, enticing the hungry reader to come and eat pork. (In North Carolina, barbecue is synonymous with pork prepared with a vinegar and red pepper sauce.) Nissenson and Jonas (1992) feature photographs of attractive and usually smiling pig images used to advertise pork products, from lard to bacon and sausage, since the turn of the century. In the mid-1950s, some North Carolina television stations featured an advertisement for bacon produced by the Frosty Morn company. This advertisement consisted of a song sung by a dancing chorus-line of happy, cute pigs: "The height of a piggy's ambition, from the day he is born, Is the hope that he will be good enough To be a Frosty Morn." In contrast, the places where pigs are produced and processed into food are hidden in the landscape.

There are economic considerations that direct the decisions of where to locate facilities that produce pigs commercially, including packing facilities. However, other factors are involved. Most late twentieth century Americans are extremely uncomfortable with the means that lead to the end of pork for the table and do not want visual or other reminders of the process and the fact that living, sentient beings are involved. A recent movie, *Babe*, portrays as hero a pig who follows his dream. It softens the production of pigs rather than depicting the actual circumstances as described by observers such as Singer (1975), Mason and Singer (1980), or Serpell (1986). The husbandry of pigs is increasingly replaced by mechanical production, with the pigs themselves machines (Mason and Singer 1980).

Efforts have been made recently to protect the environment from degradation by modern hog production (Stith and Warrick 1995; Walsh 1995; Cecelski and Kerr 1992). What is consistently overlooked are the consequences of these production systems to that part of the environment most affected: the pig itself. The movie *Babe* connects to our preferred images of pastoral tranquillity. This is how twentieth century Americans want to see this aspect of our physical environment: pigs are allowed "happy" lives, engaging in "natural" pig behaviors. While economics may indicate a mechanized system, an industrial approach to live beings is not the way that many Americans wish to envision their relationship with animals in their environment. This discomfort seems to result in the hiding of pork production from the landscape, both physically and imaginatively, while depicting pigs raised and killed for food as happy with this scenario.

Pigmalion?

How are pigs modified as a result of human perception of the animal? There are at least three ways in late twentieth century America that humans attempt to physically modify pigs, taken as material culture and resulting from the way pigs are perceived. The first is modification of pigs as agricultural commodities. Second, pigs are increasingly being viewed as a medical resource. Lastly, humans are attempting to make pigs into the ideal house pet.

The evolution of swine took millions of years. Some 9000 years ago (Serpell 1986; Porter 1993; see also the papers by Hongo and Meadow, and Redding and Rosenberg in this volume), humans began to domesticate the pig. The Eurasian wild boar (*Sus scrofa*) is believed to be the animal from which the domestic pigs of Europe and Asia descended. Curious scavengers, pigs were likely attracted to human living areas. Perhaps pigs began to accept food from humans who in turn began to confine pigs and exert increasing control over their breeding. Whatever the actual events may have been, over time the pig has become more dependent on humans (Porter 1993). As human need and the meaning given to pigs evolve, so attempts to modify the pig change.

Porcine traits do not suit the pig for nomadic life. In a sedentary agricultural setting, however, the pig's many exemplary qualities make it ideal. It can find its own food, and it can be herded or confined in pens. As an omnivorous scavenger, it can aid in cleaning human living areas. The by-product of its eating, its manure, can be used as an integral part of settled agriculture. The rooting of the pig can help to clear and control underbrush, prepare soil for tilling, rid the soil of weeds and animal pests, and remove windfall fruit that might attract various pests. This rooting behavior led to the use of pigs to find truffles in forest soils and to glean fields after harvest (Porter 1993). Pigs are food; additionally, parts of their bodies such as their fat, skin, and bristles are used by humans.

Beginning with the poultry industry, after World War II (Mason and Singer 1980) the complex, interrelated agricultural system that had to a large degree allowed animals to be animals was replaced with a more simple one. Agri*culture* became agri*business*. In this setting, the pig has one essential purpose: meat. Within this simplified agricultural setting, there is little concern about the suitability of the pig to its environment. Instead the pig, a living being, is modified to fit other human modifications of the environment. "From the moment of conception, the intensively farmed domestic pig is regulated and controlled, and rarely permitted to engage in any of the natural activities enjoyed by its wild ancestors" (Serpell 1986:7). As the greatest use of pigs through time is as food, and this implies an agricultural commodity, human efforts at modifying pigs within the agricultural setting are particularly numerous. A few examples of our re-creating pigs to fit the human modified environment will suffice.

A brief glance through Porter's *Pigs: A Handbook to the Breeds of the World* (1993) reveals a diversity of body shape and size, ear type, color, and snout structure, particularly when contrasted with the European wild boar. Body types range from rather streamlined, lean animals with relatively long legs and straight or even raised backs, to animals with dipped or sway backs, an undercarriage that nearly touches the ground, and a tendency to produce a thick layer of fat. There are large flopping ears that in some cases cover the eyes, and small erect ears. Solid colors, banded patterns, and spots in shades of black, white, and brown are seen. Asian pigs as a group are markedly different from European pigs, not to mention the pigs of other regions of the world. A map of European pig breeds in Porter (1993) shows nearly 100 major breeds, each with its own distinguishing characteristics, scattered across western and eastern Europe. The primary potential for these physical variations came from the wild boar, but humans have selectively bred pigs for the characteristics desired by humans. Nineteenth century breeders selected pigs for breeding based on growth of thick layers of fat, or lard, because of the importance of this product to humans at that time (Porter 1993). Heterosis or hybrid vigor, efficient feed to weight conversion, and quick growth are the characteristics of choice at this time in the U.S. (pers. comm. Dr. Charles Stanislaw, 1996).

No longer part of the everyday, common landscape, the pig is now produced in areas remote from concentrated human settlement. We change the appearance of our landscape to hide production of large numbers of animals that smell and create great quantities of waste products (note: it is no longer manure, a valuable resource, but "waste products"). The bucolic image of a few pigs who interact with daily regularity with a farmer is replaced with huge industrial warehouses containing hundreds of pigs, provided their necessities of life by computerized equipment. Not only are these facilities on the margins of our landscape, but humans are removed as much as possible from the process and place.

Pigs in the present-day United States are bred and genetically modified to produce large litters (so large that sows cannot nurse all the piglets) and lean meat quickly (so quickly that bone and muscle development cannot keep pace with growth) (Mason and Singer 1980). Given the opportunity, sows build a nest before giving birth to their piglets (Mason and Singer 1980), and pigs will routinely construct nests for resting and sleeping if possible (personal observation). Modern confinement systems describe sows as "farrowing units" (*Pork 1996*, January 1996), and keep the sows restrained so that they can stand, lie, eat, drink, and feed their piglets, but cannot turn around or exhibit any other behaviors (Mason and Singer 1980). Pigs warehoused in large numbers in sterile surrounds have no outlet for pig behavior or intelligence. Some pigs may attempt to relieve their frustrations by biting off the tails and chewing on the hindquarters of other pigs. The solution to these problems is not to change the pigs' environment to meet the needs of the animals, but further modify the pig for human economic and other concerns. The pigs' tails are cut off (Mason and Singer 1980; personal observation).

Stereotypical behavior, a response to a monotonous environment, is abhorred in the modern zoo, and efforts are made to construct suitable habitats and provide "normal" activities and distractions for wild animals in zoos. In contrast, pigs are physically modified to fit their environment as created by humans. As with other components of the environment, the human response to different animals—elements of the biophysical environment incorporated into the cultural environment—lies in the perceptions and meanings of the different species.

Perhaps the ultimate in the mechanization of the pig is exemplified in an advertisement featured in various pork industry publications, such as *Pork '96* and the *1996 NC Pork Report*. The ad is created by a pharmaceutical manufacturer for a parasite control medication for pigs. This product when fed to nursing sows provides protection against parasites not only for the sow but her piglets as well. The ad's banner heading reads: "Now, she can really deliver more for you" and points out that "sows are the heart of your operations' production machine." The picture, which takes well over half of this full-page ad, shows the head and body of a sow as a semi truck. Her head rests on the truck bumper and nestles between the side view mirrors and horn, fuel tank, and other accouterments of an 18-wheeler. Her body forms the trailer and cargo-carrying part of the semi. The pig is portrayed here as the ultimate machine, "à la Descartes," one could say.

There are also medical uses of the pig. Pigs provide valves for human hearts, and the primary type of insulin used by diabetics is developed from pigs. These are by-products of the slaughter of pigs for food, however, and entail using the pig essentially "as is." With the increasing sophistication of medical science and biotechnology, new ways of modifying this part of our environment emerge. Pigs are bred with a human eye disorder, retinitis pigmentosa, and are used as a model to study this disorder in people (pers. comm. Dr. Charles Stanislaw, 1996). A colony of hemophiliac pigs at the University of North Carolina at Chapel Hill is descended from pigs born in Missouri that survived and reproduced only because of continual medical and other care provided by humans. They are maintained and bred for the study of this disease in humans (pers. comm. Ms. McManus, 1996).

Companies specializing in biotechnology are attempting to develop genetically altered pigs to serve as organ donors for humans. With some 300 people dying in the United States each year while awaiting an organ transplant, the attraction of this is obvious. Called xenotransplantation, this procedure offers amelioration of the need for organ donors (Fisher 1996). The pig, the "foreigner," becomes a part of our physical body, perhaps the most intimate sharing of common space possible. Experiments in xenotransplantation began in the 1960s, and initially primates were used. In their efforts to create an animal species genetically compatible with humans for organ transplants, scientists are now focusing on the pig. There are several reasons for this shift. Limitations on the use of primates include the fact that primates can be infected with viruses that may be transmissible to humans, they reproduce slowly, and because of their intelligence and other similarities to humans, ethical considerations against using primate species are frequently raised. As Fisher continues: "Pigs can be bred rapidly and raised free of known diseases, and they have been slaughtered in service to mankind for millennia." (1996:D6).

As a last example, pigs are kept as pets. In particular, in recent years the keeping of pigs as pets, particularly as house pets, has been promoted as a fashionable activity in the U.S. The I pig of North Vietnam was imported by an exotic pet breeder on the alert for potential and lucrative pet fads (Porter 1993). Hailed as the designer pet of the '90s and the "yuppie puppy," this relatively small pig most commonly known as the Vietnamese Pot Bellied Pig, "suffered the indignity of becoming a city pet" (Porter 1993:187), and is promoted by segments of the pet industry as The (capital "T") ideal pet. Pot Bellied Pigs are described as clean, smart, having no fleas, house trainable, of a size (reportedly) to live happily in the house, and perhaps above all, a distinctive and unusual pet (Jeffery 1995; Ginsberg 1994). Like clothing and cars, a pet makes a statement about the person who owns it. Publicity for these pigs as pets almost always uses photographs of endearing little piglets, not mature adult pigs. Frequently they are portrayed wearing hats or other outfits that attempt to humanize the animals, or posed in a child's wading pool or with immaculate small children in their Sunday best.

Despite the desire of those breeding, selling, and purchasing these pigs as pets to have a mature pig weighing 40–50 pounds (or even smaller), the average weight of these pigs is about 150 pounds. The animal's intelligence requires an outlet, as do its other natural behaviors. It roots—in the carpet, linoleum, or kitchen cabinets if soil and vegetation are not available. One owner said her pig tore up the kitchen floor and ate it like fruit roll-ups (Jeffery 1995). Upon reaching puberty, territorial behaviors are exhibited, such as charging human guests visiting in the pig's house (Jeffery 1995). The pet pig often begins its life sharing the human house and home and, ironically, not infrequently shortly ends an outcast from the house and human affection. It is delivered to the local animal shelter or even slaughter house, marginal places both physically and in meaning, because its behaviors and needs are in conflict with human perception and expectation. Ultimately, these pigs are pigs, to act as they evolved over millions of years, not cuddly stuffed ornaments created by humans.

Closing Thoughts

How much of the pig is left? This essay has considered some of the ways humans view the pig, a living being in our environment, and the response to pigs as a result. The consequences of human perception and the resulting modifications of this species are considerable for the pig. The differing perceptions that twentieth century Americans have about pigs result in not only modification of the animal itself, but also the structuring of its (and consequently our) environment and where we position the pig in relationship to ourselves. It seems reasonable to ask at what point do we humans stop attempting to re-create our environment, to modify pigs and other animal species to fit our perception of that species. Should humans accept that although we can modify our environment, perhaps we should not? Should humans attempt to adapt more to the environment as it is rather than remaking it in the image that humans desire, and allow pigs to be pigs?

As humans examine the role of ethical concerns in their interactions with their environment, perhaps there is a need to move beyond consideration of the environment as "Nature" or "wildness," and consider the domesticated environment, and the individuals that are so much affected by human use of that environment. Geography's concern for human interaction with and perception of the environment, and the ways humans use space and the meanings of those uses may help us to unravel the role of our perceptions of our environment in this process of changing our world, including the animals. Geography offers a valuable perspective of human understanding and modification of other animal species, and the difficulties and consequences of these actions for our environment, including the pig.

References

Bennett, Charles F., Jr. 1960. Cultural Animal Geography: An Inviting Field of Research. *The Professional Geographer* 7(5): 12–14.

Cecelski, David, and Mary Lee Kerr. 1992. Hog Wild. *Southern Exposure* 20(3): 8–49.

Fisher, Lawrence M. 1996. Down on the Farm, a Donor:

Breeding Pigs That Can Provide Organs for Humans. *New York Times*, Business Section, 5 January 1996.

Ginsberg, Susan. 1994. Plight of the Pig. *Pet Product News* 48(7): 1, 72–73.

Hedgepeth, William. 1978. *The Hog Book*. Dolphin Books, New York.

Jeffery, Clara. 1995. Pigstown. *Washington* [DC] *City Paper*, 21 July 1995, pp. 18–30.

Kearney, Milo. 1991. *The Role of Swine Symbolism in Medieval Culture: blanc sanglier*. Edwin Mellen Press, Lewiston, NY.

Lowenthal, David. 1968. The American Scene. *Geographical Review* 58(1): 61–88.

Mason, Jim, and Peter Singer. 1980. *Animal Factories*. Crown Publishers, New York.

Nissenson, Marilyn, and Susan Jonas. 1992. *The Ubiquitous Pig*. Abradale Press, New York.

NC Farm Bureau News, May 1995.

NC Pork Report, January 1996.

Pork '96, January 1996.

Porter, Valerie. 1993. *Pigs: A Handbook to the Breeds of the World*. Comstock Publishing Associates (division of Cornell University Press), Ithaca, NY.

Regan, Tom, and Peter Singer (eds.). 1976. *Animal Rights and Human Obligations*. Prentice-Hall, Englewood Cliffs, NJ.

Relph, Edward. 1981. *Rational Landscapes and Humanistic Geography*. Barnes and Noble Books, Totowa, NJ.

Serpell, James. 1986. *In the Company of Animals*. Cambridge University Press, New York.

Sillar, F.C., and R.M. Meyler. 1961. *The Symbolic Pig*. Oliver and Boyd, London.

Simoons, Frederick J. 1974. Contemporary Research Themes in the Cultural Geography of Domesticated Animals. *Geographical Review* 64:557–576.

Singer, Peter. 1975. *Animal Liberation*. Avon Books, New York.

Stith, Pat, and Joby Warrick. 1995. Boss Hog: North Carolina's Pork Revolution. *News and Observer* [Raleigh, NC], 19–26 February 1995.

Tuan, Yi-Fu. 1984. *Dominance and Affection: The Making of Pets*. Yale University Press, New Haven.

Vale, Thomas R., and Albert J. Parker. 1980. Biogeography: Research Opportunities for Geographers. *The Professional Geographer* 32(2): 149–157.

Veblen, Thomas T. 1989. Biogeography. In *Geography in America*, ed. G.L. Gaile and C. J. Willmott, pp. 28–46. Merrill Publishing Company, Columbus, OH.

Walsh, Bill. 1995. Waste is a Terrible Thing to Mind. *Cooperative Farmer* (October–November): 17–21.

PRIVY-PIGS IN PREHISTORY?
A KOREAN ANALOG FOR NEOLITHIC CHINESE
SUBSISTENCE PRACTICES

David J. Nemeth

Department of Geography and Planning, University of Toledo, Toledo, Ohio 43606

Unlike many of our other domesticated animals, pigs do not have an important dual role, the pig being kept almost solely as a meat producer. (Heiser 1973:58)

Pigs-as-Pork

Do pigs have a past? The theme of this book—"pigs in prehistory"—seems to affirm pig life simply by asserting it. For philosophical reasons I will pursue a seemingly absurd question: "Were there pigs in prehistory?" Common sense emphatically answers "Yes!" but I will additionally turn to the science of archaeology in search of the usual kinds of validity and authenticity that clarify the meaning of pigs for the purposes of carrying on a rational discussion about them.

When 10,000-year-old pig bones were found a few years ago at Hallan Çemi, in Turkey, a *New York Times* interview with archaeologist Richard Redding (see Redding and Rosenberg paper in this volume) reported that the discovery "strongly suggests that the pig was the earliest animal that people domesticated for food" (Wilford 1994:B5). Note here, as in the epigraph, that the scientific story of the existence of pigs in prehistory is accepted as axiom, while the role of the pig as a source of nutrition in subsistence is accepted as its corollary. While it would seem to be a simple task to separate the two stories for purposes of clarification, my point is that *in practice* the two stories converge as one truth, and that they have been widely disseminated as such in both the scientific literature and the popular press.

One does not have to turn to the *New York Times* to demonstrate that the in-place habits of everyday human thoughts and belief regarding pigs are more "in the present" than "in the past." If the present cultural conversation of humankind on pigs as a source of nutrition is pork-centered, its emphasis on pig-as-pork also projects itself into the past in the archaeological literature. Here is a statement from a typical archaeological conversation that simultaneously affirms the importance of pigs in prehistory and presupposes their importance "for food": "The relationship between humans and pigs is biologically symbiotic in that humans provide pigs with fodder and pigs provide humans with nutrition and a medium for symbolic activities" (S.O. Kim 1994:132).

What is the "truthfulness" of an assertion that reduces the essence of pig life in the past to providing "humans with nutrition" (and infers symbolic activities from this emphasis)? This paper challenges the validity and authenticity of the statement by revealing the extent to which current archaeological conversation on the topic of pigs in prehistory presupposes and exaggerates its importance as pork. It also explores how the exaggeration of pigs-as-pork in prehistory limits the archaeological discussion about the richness of pig life.

Modern everyday conversation on pigs in a global economy is pork-centered to the extent that its silence on the richness of pig life is sometimes deafening (Fig. 1). Any ambiguity in the story about which pigs "go to market" and which "stay home" has been reduced to a certainty by the time that American children reach adulthood—all pigs go to market. A typical encyclopedia entry on pigs identifies them as "lard, bacon, meat, or pork types" (Wilson 1996:87). Within these narrow limits of adult conversation on pigs-as-pork, the richness of pig life has been lost. Pig life is likewise a marginalized topic in academic forums where the social reproduction of pigs-as-pork in scholarly exchanges also deprives pigs of any meaningful existence or importance other than as pork. As we shall see, archaeology also participates in

Fig. 1. Pig transport by dirt-loader. A pig slaughter in a Taiwanese factory-farm during a recent hoof-and-mouth disease epidemic. Modern factory hog confinement systems found worldwide can each concentrate tens of thousands of pigs in metal barns with concrete floors, to eat, excrete, and reproduce in unhealthy alienation from a natural biological cycle (Anonymous 1998). The press interprets this epidemic as a global economic disaster for the pork industry, rather than as an ethical issue (Tyler 1997).
Drawing by James Ashley

this contemporary cultural conversation about pigs in which pigs have no presence.

In modernity's conversation, pigs live mainly in the shadow of their shelf life. Any meaningful discussion on pig life is perpetually postponed by the urgencies of discussions about increasing human food needs and economic expectations. Consider, for example, the irony, paradox, and tragedy in an everyday phrase like "pork futures," where the importance of pig life in the present is trivialized by predictions for profits after pig death. Ironically and paradoxically, pigs have no futures within the constraints of the dominant pork-centered discourse about them. Tragically, by limiting the present conversation on pig/human relations to the economics of pork production, transportation, and consumption, humans constrain the potential of their own compassion for constructing a better quality of life for both humans and pigs in the future.

Three Kinds of Archaeological Stories About Pigs in Prehistory

This paper is organized around a distinction between three kinds of archaeological stories about pigs in prehistory: (1) archaeological stories that "visualize" the evidence of pigs in prehistory, e.g., straightforward, descriptive stories—scientific reports—that strive to let the evidence "speak for itself"; (2) archaeological stories that go beyond the visible evidence to tell a story about pigs in prehistory; e.g., interpretive stories with literary style and panache that deploy analogy to tell a tale about the past terms of the present; and (3) archaeological stories that use the visible evidence as an opportunity to tell a story about pigs in prehistory that is mainly a story about conditions in the present, e.g., political stories that explore the potential of a critical archaeology to challenge the exclusive knowledge claims of a privileged discourse. Discerning among these kinds of archaeological stories can offer some insight into the process by which an exaggerated archaeological conversation on the importance of pigs-as-pork in prehistory (pigs go to market) becomes privileged over an alternative story, for example, pigs as fertilizer factories (pigs at home).

An interesting exchange of ideas among scholars on the topic of pigs in prehistory was recently published in *Current Anthropology* (1994:119–141; 1995:292–293). One participant in the *Current Anthropology* (hereafter *CA*) exchange recommends using ethnographic analogy from South Korea in an effort to expand archaeological discussion on the roles of pigs in prehistoric subsistence (Nelson 1994:135–136). Nelson specifically mentions coprophagous privy-pigs as sources of human nutrition. Although she does not challenge the privileged story of pigs-as-pork in prehistory, her evoking the possibility of privy-pigs as food can provoke a deep visceral reaction in people who find comfort in assuming that the boundary zone between eating and defecating is natural, secure, and certain rather than socially constructed, permeable, and ambiguous.

This paper elaborates on Nelson's story of Korean privy-pigs as an ethnographic analogy that can offer persuasive evidence in support of extending pig importance in prehistory to aspects of pig life other than as pork, for example, by constructing a detailed case for privy-pigs as fertilizer factories and as disease-control mechanisms in subsistence agriculture. As indicated, the pigsty-privy as an ethnographic analogy can be both informative and disruptive. The paper concludes that a disruptive, critical archaeology can transcend more familiar archaeologies of substance and style to reveal the limits of their truth-seeking claims, for example, by revealing the exaggerated importance of pigs-as-pork in prehistory.

Visualizing the Importance of Pigs in Prehistory

The *CA* forum on pigs in prehistory centered on an innovative research article by Kim Seung-Og titled "Burials, Pigs, and Political Prestige in Neolithic China" (1994). Kim reports that pig bones are ubiquitous discoveries in Chinese archaeological excavations from Neolithic settlements and tombs. He notes, however, that "Archaeological research in China has been fragmentary and mostly concerned with chronology and typology" (1994:122). He adds that this research has not been systematically applied to the scientific study of prehistoric social organization, its complexity, and its evolution. Kim argues for an application of this archaeological data, now narrowly framed as a fragmentary visual reconstruction of archaeological evidence at scattered excavation sites. He provides his own interpretative extrapolation from the visible evidence by telling his story about the significance of pigs in prehistoric social evolution. He deploys ethnographic analogy to interpret some of the visible archaeological data involving pigs in a region of Neolithic China as a hypothesis (his story) about the growth of political complexity there.

Kim's article reproduces site plans for (and tables of artifact typologies based on) archaeological excavations in Shandong that represent a chronology covering four periods, from 4300–1900 B.C. It is important to distinguish what Kim has to say about the archaeological evidence from the spatial configurations of the visible site plans themselves; for instance, the visible excavated space of Shandong's Neolithic settlements (site maps) can be contrasted to what Kim enunciates (his story, his discourse) about the political role of pigs in prehistory and in subsistence. While Kim's story and the visible evidence impinge on each other, and while pigs are objects of knowledge in both the visible and the sayable, each has its own separate history of formation. The visible spaces and the stories they motivate are each separate knowledge domains having different histories and modes of organization. They are shaped by different forces.

Thus the site maps of Neolithic Shandong, their content, spaces and voids, speak for themselves but don't say very much. In contrast, Kim has a lot to say about pigs in prehistory based on the shape of the visible evidence. However, his story is not shaped in the past, but in the present. His novel political thesis impinges on the visible site plans but presumes the special significance of pigs-as-food in Chinese culture.

Ethnographic Analogy: Pigs in Prehistory (in Terms of Pigs in the Present)

Kim promotes ethnographic analogy "as a source of plausible hypotheses which can be tested against the archaeological record" (1994:138). His analogy begins by extrapolating from present conditions into the past. He points out the long history of pig domestication in China, and that "Pigs were the predominant domesticated animal in the Neolithic as they are today" (1994:119). He notes that

> In contemporary China,...the pig is overwhelmingly the dominant source of the protein and fat. Therefore, it is not surprising to find that China has one of the largest sausage and ham industries in the world. The ups and downs of the Chinese economy are estimated in terms of the availability and quality of pork. (1994:121)

Kim also cites numerous ethnographic studies that show that pigs make and have made a major contribution to human diet and symbolic feasts throughout Asia, such as in Melanesia (see also Oliver et al. 1993).

Kim proceeds to argue that archaeological studies of Neolithic burials in Shandong, China, demonstrate that intensive pig production was important not only for human diet, but for ritual, and also for the display of individual wealth and inequality in the rise of political elites: e.g., "pigs functioned not only as means of subsistence (e.g., protein) but also of objects of social production (e.g., ceremony, feasts, status display, exchanges, etc.)" (1994:132). Thus, Kim's conclusions about the sociopolitical implications of pig control in the emergence of complex societies are mainly based on the assumption that pigs "were of critical dietary and ritual importance" (1994:120).

Kim argues that "except for pigs and dogs most animals were quantitatively of little importance to subsistence" and that "In contrast to dogs, which are found mainly in residential areas, domesticated pigs are abundant in both settlements and burial sites, which leads to the speculation that only pigs were given symbolic treatment. The small circular pits found in great numbers near the settlements usually contained complete pig skeletons...and it seems reasonable to hypothesize that they functioned as structures designed for raising pigs" (1994:123). Kim identifies "Almost all of the pig remains in...mortuary contexts [as] skulls or jaws. Pig figurines are also not uncommon" (1994:123). Unfortunately, his article contains no descriptions or photographs of these pig figurines.

At one point in his article Kim misses a chance to use ethnographic analogy to broaden the discussion of the role of the pig in subsistence beyond food, when he cites Rappaport's finding (1967:121) that pigs are "very resistant to disease and produce a large amount of fertilizer for farming." Fortunately, a follow-up comment by Sarah M. Nelson raises the possibility that Kim's "small circular pits" (she calls them "small round houses") with complete pig skeletons might have something to do with "the association of pigs and latrines in both Korea and China" (1994:136). She wonders if this practice might not have had "a very long tradition" in East Asia, then concludes that "It makes good ecological sense for pigs to process human excrement and turn it into meat that humans can consume" (1994:136).

While Nelson and the other commentators offer a range of challenges to Kim's political thesis, no commentary challenges his underlying assumption about the paramount nutritional role of the pig in subsistence agriculture. Yet Kim constructs his entire political thesis from the premise that the "small circular pits" are holding pens for concentrating pigs-as-pork. He makes no reference to privy-pig systems in his ethnographic analogies and, although he acknowledges that pigs produce fertilizer for other subsistence systems, he does not interpret the pits as fertilizer factories in Neolithic China. Commentators seem to be in casual agreement with Kim's assumption about the primary importance of pigs-as-pork when they write of "relations between givers and recipients of the distributed pork" (Rosman 1994:137); or that "Pigs were amenable to social manipulation because they were such a valuable food resource" (Quilter 1994: 137); or that "One possible way to use pigs for long-distance exchange would have been to convert the pork into ham" (Lee 1994:135). Thus many of the commentators also agree to presume the importance of pig-as-pork in Neolithic subsistence. Even Nelson's expansive line of questioning about the possibility of diverse roles for pigs in subsistence supports Kim's presumption about the paramount significance of the pig in Neolithic China as a provider of protein and fat for human nutrition. Her commentary is even accompanied by a photograph of a large pig-going-to-market on the back of a bicycle.

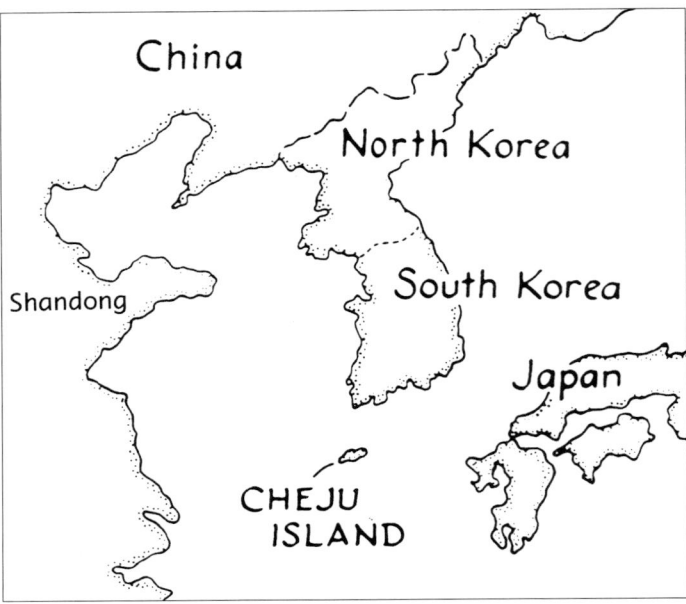

Fig. 2: Site and situation: Cheju Island, South Korea.

There is clearly a consensus of opinion in this scholarly exchange as to the importance of pigs in prehistory. One commentator even writes "The fact that pigs were important in ancient and modern China, however, does not appear to be an issue that needs a great deal of argument to justify" (Quilter 1994:137). However, the assumption that the importance of pigs in prehistory was as pork does seem to require further justification. This is especially so since Kim's rejoinder to his commentators makes no reference to the implications of Nelson's suggestive ethnographic analogy about "the association of pigs and latrines in both Korea and China" (1994:136).

In sum, both Kim, Nelson, and some other commentators deploy ethnographic analogy in their arguments. They tell analogous stories about pigs in prehistory in terms of the importance of pigs in the present; for example, they bring the ethnographic cases from Melanesia and Korea to bear on the archaeological case from China. The following section responds to Nelson's suggestion that the association of pigs and latrines in Korea be explored in more detail.

Pigsty-Privies on Cheju Island (South Korea)

Site and Situation

Cheju Island is located in the East China Sea several hundred kilometers southeast of the Shandong peninsula (Fig. 2). Cheju Island was called "Yongju" in early

Chinese dynastic times. Histories report that the "First Emperor" (ca. 230 B.C.) sent young men and maidens into the East China Sea with "the five grains" in search of Yongju and its plants of immortality. Local Cheju Island legend corroborates these events, suggesting that ancient Chinese subsistence practices, including pigsty-privies, probably arrived in Cheju Island at an early date. Indirect routes of diffusion from China through the Korean peninsula or from points south in the East China Sea like the Ryukyu Islands are also possible.

Subsistence feeding habits of people and pigs on Cheju Island were well-integrated in the pigsty-privy system. Most Cheju farming villages are at or near the coast. Islander survival depended upon a marginally productive low-technology farming system that integrated pig raising with millet growing, supplemented by the hunting and gathering of nearshore ocean resources (for example, shellfish and kelp). The earliest description of the Cheju Island living environment is provided by the Chinese: "The people follow humble customs. They wear boarskins. In summer they live in leather houses. In winter they live in caves. There are five crops grown there. They use iron teeth for digging in the earth" (*T'angso*, ca. A.D. 661–663).

In an early twentieth-century account, the American adventurer Malcolm Anderson observed that the pigsty-privy was "most important" to the Cheju Islanders (1914:396). We can infer that the custom probably existed on the island for many centuries prior to his report. It is also significant to note that all early Western travelers described Cheju Island as a primitive yet productive agroecosystem, and when Chaille-Long visited Cheju in the late nineteenth century, he remarked on "the good physique of both the male and female" (1890:249). This indicates that Cheju Islanders prior to modernization were a robust population. Was this because of, or in spite of, their pigsty-privy customs?

Izumi (1971:21) reported that in 1934 over 90% of Cheju Islanders were from working farm families, and as recently as 1963 farm households still constituted over 85% of the island population. In 1934 there were approximately 200,000 Cheju Islanders, comprising about 35,000 farm families, mostly living in 500 villages scattered on or near the island's coast. Each farm family had on average one pigsty-privy and one or two pigs. Thus, until fairly recently, pigsty-privies were part of an enduring subsistence folkway on Cheju Island.

Academic debate arose during the 1970s over the virtues and vices of the pigsty-privy system on Cheju Island. This debate was quickly rendered moot when the government launched a successful campaign in the early 1980s to eradicate the practice of privy-pig keeping on the island. The government's campaign was initiated ostensibly to promote human and pig health and hygiene on the island. Government economic planners in Seoul also considered Cheju's pigsty-privies to be offensive to the sensibilities of international tourists. They were also embarrassing to the reputation of South Korea as a rapidly developing country.

Elaborating on the benefits of privy-pig keeping on Cheju Island helps put the presumed primacy of the nutritional factor of subsistence pig-keeping in clearer perspective. In the main, Cheju Island's privy-pigs clearly played important roles in subsistence as fertilizer factories, and probably also in disease control. I came to appreciate these possibilities during approximately six years of residency and field work on Cheju Island (1973–1974; 1980–1981; 1984–1985; 1995). In addition to personal experience, I collected and analyzed locally published survey materials on the pigsty-privy system, written mostly by Korean medical, agricultural, and social scientists. The following account is a pastiche of

Fig. 3. Photograph of an abandoned pigsty-privy in a "folk village," or "living-museum" on Cheju Island. A sign posted for the benefit of tourists reads: "This is a combined pigpen and outhouse that our ancestors built. Its appearance is preserved here, but it is no longer used."

findings excerpted from some of my previously published work (1985; 1987; 1989; 1995). They are revised and reintroduced here mainly for the purposes of ethnographic analogy in the context of a philosophical and methodological discussion. I prefer to defer to the clarity of my memories from the early 1970s on Cheju Island, and to speak of the pigsty-privy system in the following passages as if it still existed (Fig. 3).

Pigsty-Privy Morphology and Layout

Pigsty-privy structures on Cheju Island are of various shapes and sizes, but their basic morphology and layout combines a swine pen and a human latrine into a single structure (Figs. 4, 5 and 6). Park and Chyu (1963:161) report on the widespread distribution of such structures in traditional Asian agroecosystems. Their invention is Chinese. The Chinese ideogram 圂 signifies both "pigsty" and "privy" and is pronounced "hon" in Korean. It depicts a pig 豕 within an enclosure ☐.

Terracotta images of pigsty-privy structures have been excavated from Han Dynasty (207 B.C.–A.D. 220) tombs (Laufer 1962:53; Hsu 1980:98; Bray 1984:291). The fact that these grave models (Figs. 7 and 8) were buried with kings indicates the importance of the pigsty-privy structures at an early time in East Asia. We might recall here that Kim mentions "pig figurines" among mortuary evidence excavated from Chinese Neolithic tombs in Shandong. Simoons (1961:27) notes in general the continued importance of the pig in the traditional economy of East Asia into the modern era, as "a household scavenger…quartered in the family garbage pit and regularly fed human excreta and garbage." Examples of the privy-pig waste-recycling system persisting into the twentieth century include North Korea (anecdotal) and also the Ryukyu Islands (Glacken 1955:71; Pitts, Lebra, and Suttles 1955:190).

The morphology of the Cheju Island pigsty-privy in its simplest and most primitive form consists of an elevated perch—two footboards (or stone slabs) on which one squats while defecating—positioned above the enclosed pit of a pig-pen. The distance between the perch and the pit below reflects a safety factor necessitated by the sometimes threatening enthusiasm of the voracious, coprophagous privy-pig. On Cheju, the pigsty walls are invariably made of piled angular basalts, constructed to a height that prevents the escape of the pig. In 1984 I interviewed a young man named Kim Jun-ho from Mara Islet, near Cheju Island. He talked briefly about privy-pigs and pigsty-privy construction:

> The pigs are small and lean… Islanders construct their pig pens with stones. The area and shape of the pen varies from house to house. The wall they build to form a pen is about half-a-meter or so in height. They dig the ground one or two meters deep inside the pen. At the corners of the pen they install a shelter for the pig. A secure perch to support a defecating person is then installed high above the pen.

By convention, farm family members of all ages and both sexes use the privy and therefore do not normally defecate indiscriminately in their living environment. Concern for modesty and the need for some protection from winter winds usually leads to the construction of a walled privy closet enclosing the squatting perch. This is also constructed of piled basalts. This temporary human retreat may be waist high, or even high enough to permit one's standing within. The privy roof may be covered with straw and weighted down with rocks or even tied down to prevent its being blown away by Cheju's notorious high winds. The pig is invariably provided with some kind of covered shelter, too, usually at the opposite end of the pig pen from the privy. The privy basement has open access from the pig pen. A stone feeding bowl in the pen is the only other item typical of the island's pigsty-privy structures.

Fig. 4. Cheju Island pigsty-privy, 1981.

Fig. 5. Cheju Island pigsty-privy, 1981.

Fig. 6. Cheju Island pigsty-privy, 1981.

The location of the pigsty-privy within the farm compound varies, but typically is adjacent to, and to the rear of, the main dwelling unit; this is for the sake of human convenience, since the trip from house to privy is often made in severe weather. However, for superstitious or aesthetic reasons the pigsty-privy may be located further away. For example, geomantic (Korean: *p'ung-su*) theory and practice may have influenced the location of pigsty-privy structures in the farm compound.

The walled pigsty may have been excavated to a depth of several feet when constructed in order to facilitate the concentration, collection, and storage of organic waste materials. The manufacture of compost from household wastes is one of the principle virtues of the pigsty-privy system. Composting reduces fly and rodent problems. It also produces a rich manure that is odorless and largely free of disease organisms (Oelhal 1978:218).

The Pigsty-Privy as a Fertilizer Factory

From a functional perspective, the pig's major role on Cheju is as a "fertilizer factory" producing dung that is rich in nitrogen, phosphorus, and potassium for composting. The compost will be used to enhance Cheju's mineral-starved soil, enabling continuous agricultural production. Humans contribute household organic wastes and their own body wastes as pig-feed. Park and Chyu (1963:166) estimate that human feces constitute as much as one-third of the privy-pig's diet. Grain hulls form the second major ingredient of the privy-pig's feed, supplemented by fish viscera and other organic wastes. Human urine, which is particularly high in nitrogen and pathogen-free, may sometimes be collected separately for use as a garden fertilizer in vegetable patches. Human feces, on the other hand, are inappropriate for enhancing Cheju's acidic volcanic soils. They are also infested with disease-causing organisms, like the eggs of parasites, and are not so easily and safely disposed of in the farm environment. Flies are especially attracted to human feces and can spread infection and disease as they move freely about the farmstead environment, indoors and outside. Reducing flies reduces the spread of disease on the farm. Thus the pigsty-privy system may also be discussed as an effective method of disposing of dangerous human wastes and reducing flies.

To aid the composting process in the pig pen, the farmer occasionally adds millet, barley, and rice stalks. These have a high carbon content and will serve as bedding litter and absorb the pig's wet dung and urine. The pig normally eats warm human feces on deposit, so this item is not part of the material normally undergoing biodegeneration in the compost pit. Where there is an active pig, pathogenic organisms that may inhabit human feces have little time to disperse into and contaminate the environment of the farm compound. The daily antics of the pig, criss-crossing the pen from shelter to privy and back again, serve to aerate the composting materials and accelerate the decomposition of all the organic wastes. The compost is removed from the pen once or twice yearly, and transported to the fields as manure. This process of waste recycling and manure manufacture in the pigsty-privy is described and explained in more detail in Nemeth 1987 (pp. 177–180).

The Role of the Pigsty-Privy in Disease Control

This interesting aspect of the pigsty-privy system is highly provocative, not widely discussed, and deserving of a detailed description in so far as it may also relate to the importance of pigs in contributing to the robustness of Neolithic Chinese subsistence farmers. Arguing the positive role of the

Fig. 7. Clay figurine of combined pigsty and privy excavated from a Han dynasty tomb. When Laufer first published this illustration in 1909 he misidentified the construction as a "grain-tower" for fattening pigs (Laufer 1962:53). Bray corrected this error when she republished the illustration and called the construction a "combined pigsty and privy" (1984:291).

Fig. 8. Terracotta image of combined pigsty and privy excavated from a Han dynasty tomb. Drawing by James Ashley

pigsty-privy system in subsistence disease control directly conflicts with the South Korean government's major rationale for its privy-pig eradication campaign. But then, one must wonder why such an efficient, inventive, and productive agricultural civilization as China's, which provided its people with "the highest standard of living in the world" as recently as the eighteenth century (Spencer 1954:315), would invent pigsty-privies and bury clay miniatures of them with their emperors if pigsty-privies endangered rather than enhanced the public good?

Cheju Island's pigsty-privy system seems to defeat the survival strategies of some human endoparasites (e.g., roundworm, hookworm, and whipworm). Ironically, it also ensures that one particular endoparasite flourishes. This parasite is the *Taenia solium* tapeworm that infects almost all Cheju Islanders. *Taenia solium* is called "the pork tapeworm of humans" because (1) the human is its only definitive (primary) host, and (2) the pig is its intermediate (secondary) host.

Those primers on parasites that focus on flatworms reveal that *T. solium* depends on a continuing interaction between pigs and humans, both active and highly mobile free-living animals in the environment. The unconfined environment of these animals is, however, deadly to the mature *T. solium* organism that can survive only within its human host. Since the *T. solium* endoparasite is confined to the intestinal tract of its host, and since it does not produce progeny within its confinement nor can it safely venture outside to do so, its survival strategy depends on its eggs somehow dispersing in the free environment and reaching the digestive tract of the active intermediate host—the pig. Once within the pig, these eggs can then hatch and develop into the tapeworm's immature phase of encysted bladderworms (cysticerci). These bladderworms must eventually reenter the intestinal tract of the definitive host—the human—normally by ingestion, in order to mature into adult *T. solium* organisms.

Relatively few *T. solium* eggs are able to survive the array of hostile environments they encounter on the way to becoming mature bladderworms and, finally, adult tapeworms. In the beginning of its life cycle, the adult *T. solium* endoparasite attaches itself by means of assorted hooks and holdfasts to the inner walls of the human small intestine. The gut of the infected host may also harbor other animal life, such as *Taenia* organisms of several species, and various other kinds of helminthic parasites—roundworms, hookworms, whipworms, and others. These, in addition to lung flukes, amoebae, and other numerous non-helminthic parasites can coexist, and together they will sap their host's nourishment and strength. Hookworm infections, for example, can result in anemia, mental and physical retardation in children, and apathy toward work and other physical exertion. The infected human host, undernourished and weakened by parasites, may become more susceptible to the potential danger presented by any of its individual endoparasites, and less resistant to all manner of disease.

The *Taenia* organism has a long, flat, segmented body that normally streams away from its anchored head (scolex) through the alimentary tract and toward the host's anus. The animal grows in length by adding segments, called proglottids. The young proglottids constantly form close to the scolex; the oldest proglottids are found farthest away from this formative region. A typical adult tapeworm is about 700 to 1000 proglottids in length (or about 3 m). The constantly forming proglottids will eventually detach, oldest first, from the tail of the adult worms, in groups of five or six segments. These will enter the free environment in human fecal deposits.

Formative proglottids of *T. solium* undergo hermaphroditic reproduction on their maturity, and each may carry 40,000 eggs (onchospheres) out of the host and into the free environment. In the free environment *T. solium* eggs are relatively durable and become widely dispersed. Some of them will be picked up and will infect the parasite's intermediate host, the pig. If the proglottid is not quickly consumed, it will disintegrate and release its eggs to be dispersed by wind, water, and other forms of natural transport.

Holmes (1976:30–31) reports that gravid (egg-laden) *T. solium* proglottids are motile and can migrate away from the fecal deposits of the primary host. The *T. solium* organism relies upon activities and movements of its two hosts in the free environment that are beyond its control, but are not in fact entirely random. Through the motility of its proglottids, the tapeworm increases the chances of infecting its intermediate host (pig) with its eggs by maximizing its chances of survival according to the probable feeding habits of the pig. While the pig is omnivorous and coprophagous, it is not necessarily fond of fresh human feces. If the pig normally avoids human feces, this helps explain why tapeworm proglottids have evolved to move clear of the feces.

Schmidt and Roberts report (1981:378) that *T. solium* must overcome tremendous odds to successfully infect its intermediate host (pig) in the free environment. The tapeworm bombards the free environment with massive doses of its eggs in order to maximize the odds of a productive encounter. The primary host (the human) of course contributes to the cause of *T. solium* by indiscriminately defecating here and there in the living space shared by man and pig. The proglottids disintegrate rapidly and release the onchospheres. Each egg may remain viable for some weeks, but exposure to direct sunlight and to extremes of heat or cold adversely affect viability.

As *T. solium*'s intermediate host, the pig provides a safe haven for harboring the immature parasite, whatever happens to the mature *Taenia* (Holmes 1976:27). Even should all adult worms inhabiting their human hosts in the vicinity be exterminated, cysticerci-infected pigs provide a reservoir for a future generation of infective mature tapeworms.

To complete *T. solium*'s successful life cycle, a *Taenia* egg must be ingested by a pig. Within the pig's alimentary tract the embryophore surrounding the egg will dissolve, releasing an embryo capable of penetrating the wall of the small intestine to enter the peripheral circulatory system of the pig. From there, the embryo enters the muscular tissue, or other flesh organs of the pig, and begins to metamorphose into an infective bladderworm (cysticercus).

The cysticercus develops, over a 60-day period, into a pea-sized transparent ellipsoid. The pig's tongue, neck, and shoulders are typically the most heavily infected parts, though no edible part of the pig is completely clear of cysticercus infection. Since the cysticerci in infected pork are quite large, and therefore visible, humans can usually detect the danger and avoid eating them. However, visual inspection of pork is not foolproof. Sometimes the bladderworm is too small, too isolated, or too deeply nested in the pork tissue to be detected. Humans may still avoid infection by cooking the infected meat thoroughly and killing the cysticercus.

It is important at this juncture to distinguish between the health implications of adult *Taenia* infection in humans, and encysted bladderworm (cysticercus) disease in humans. Schmidt and Roberts (1981:377) report that "the most potentially dangerous adult tapeworm of humans is the pork tapeworm, *Taenia solium*, because of the possibilities of self-infection with cysticerci. Furthermore, it is possible to infect others in the same household with the juveniles of this parasite, often with grave results." Thus, we can speak of two distinct paths in the life cycle of *T. solium*: the "normal" path involving the pig as intermediate host and the "aberrant" path where *Taenia* eggs infect humans directly. Should a viable *T. solium* egg or embryo become accidentally ingested or otherwise reach the upper digestive tract of a human, then bladderworms (cysticerci) would begin to form within the human tissues. This growth of the encysted pea-sized bladderworms in man—the disease is called cysticercosis—can eventually cause localized pain in the infected skeletal muscles. Worse, if the cysticerci become located in the eye orbits or in the brain, they can cause blindness and epilepsy (Hegner et al. 1938:341). Until recently, the disease was inoperable in the brain.

Accidental infection of humans with *T. solium* eggs and embryos is also feasible should the tail (strobilia) of the adult worm reverse its direction in the upper intestinal tract and somehow invade the stomach, releasing viable proglottids there. A more likely accident involves the ingestion by the definitive host of a fugitive *T. solium* egg or embryo somewhere in the free environment, for example, by its feeding on egg-contaminated vegetables or drinking water. Presumably, the pigsty-privy custom serves to minimize the second kind of accident by restricting potentially contaminated human fecal deposits to one small part of the free environment (a pigsty-privy) for a very short period of time, until a privy-pig consumes them. In order to better understand the pigsty-privy as a solution to a human survival problem that involves pigs and pork tapeworms, it is useful to trace the evolving interactions of human, pig, and tapeworm in the subsistence environment.

The implication thus far is that a good quality of life was possible despite the near-hundred-percent certainty of *T. solium* infection among the islanders, and that the pigsty-privy may have played an important role in modifying human behavior in a significant way toward creating and maintaining that productive and healthy environment in spite of the *Taenia* organism.

Foodways of Cheju Islanders

Cheju Islanders have been known to exuberate over the flavor of pork from their privy-pigs (Park and Chyu 1963:168). Pork, the most popular flesh food for contemporary Cheju Islanders, was likely the most-craved during subsistence times, too, especially so since it was consumed only infrequently. Yet there is no evidence that pig-as-pork was ever the primary importance of the pig in Cheju Island subsistence. Fish, shellfish, and many other tidal and nearshore aquatic creatures were not only abundant, but their exploitation was assured by a centuries-old "female diver" tradition on the island. There were also chickens. Beef consumption has increased dramatically over the past several decades.

Village celebrations have traditionally provided opportunities for Cheju Islanders to slaughter and consume their privy-pigs. This does not mean there was a lot of pork available for consumption. Not only were privy-pigs few and dispersed in subsistence villages, the breed of privy-pig used was very small.[1]

The pig-eating celebration, which is largely a male activity and often accompanied by liquor, is called *churyum*. The usual pork morsels consumed at a *churyum* are from the viscera, including stomach, liver, and other parts. These parts are eaten raw. Contemporary Cheju raw-pork eaters are aware of the danger of pork tapeworm infection in their environment. They realize it is dangerous to eat "measly" (cysticerci infected) pork and can refuse to do so. Of course, the island could be freed from pork tapeworm infection and cysticercosis if the morsels were always cooked thoroughly before being eaten. The heat necessary to cook the pink out of the pork kills the cysticerci. But this is not the eating habit of the Cheju Islander.

Islanders prefer to believe instead that not drinking cold water after eating raw pork prevents their infection with tapeworm (Park and Chyu 1968:168). If they discover that they are infected with an adult tapeworm, there are several natural remedies found in their environment to kill the parasite; for example, torreya nuts. Perhaps a more satisfactory precaution against human pork tapeworm infection is taken with regard to the way privy-pigs are raised on the island, and the age at which they are selected for slaughter. Islanders have determined that sucklings are most vulnerable to infection while in the privy-pigsty; they are therefore removed from the pens until they are older and have developed some of their natural immunities to parasite infection.

The high level of awareness among the islanders regarding the causal relationship between the human tapeworm disease and the cysticercus-infected pork is disputed in a survey by Han, Kim, and Kang (1971). Their survey of 70 Cheju farm households indicated that half of those interviewed were ignorant of the existence of cysticercus infection and its danger. While it may be true that human cysticercosis is unknown to them, they are certainly aware of pork cysticercus infection, for they call the pea-sized bladderworms they discover in their pork by the term *nangch'ung*. Their own pork tapeworms they call *choch'ung*. The common suffix *ch'ung* in the two words means "bug": *choch'ung* means "bug-in-a-line," describing the proglottid segments, or the strobilia of adult tapeworms; *nangch'ung* means "pocket bug," or the pea-sized cysticercus in the infected pork. The suffix of *ch'ung* in both words perhaps indicates that the islanders are aware of the relationship between the *T. solium* adult and immature forms in their environment.

With regard to the cuts of pork actually consumed, honored guests at *churyum* celebrations are most likely to receive liver, stomach, and fetus, in that order, followed by other viscera (uterus, brain, intestine, pancreas, and kidney). Skeletal muscles of the pig—which constitute the greater part of the pig's edible parts—are not a preferred portion, but are widely consumed by the majority of the raw pork eaters. Most of the low-rank pork-eating islanders need the protein and fat, but are not honored enough to receive the most favored portions. Pork tongue is the least preferred morsel, but is nevertheless eaten by someone.

The tongue is the morsel most likely to be infected with the *T. solium* cysticerci, followed by the neck, shoulder, and other skeletal muscles, then the various viscera, and finally—the organ least likely to be infected—the liver. As a patriarchal ritual, consumption of privy-pigs for *churyum* tends to favor the health and survival of the dominant males in this island society. While avoiding tapeworm infection seems impossible, especially for lower-ranking males in Cheju society, the pigsty-privy system works to minimize both their infections from other endoparasites and the chance of cysticercosis disease among them.

Island women are also infected by pork tapeworms. Mudang (shaman) rituals and other special occasions involving women are often accompanied by privy-pig consumption. In general, Cheju womenfolk have a closer physical proximity to the privy-pigs on a day-to-day basis than do men. This is in part because pig raising has been traditionally women's work (Cho 1979:43).

Presumably, the severity of *Taenia* infection and cysticercosis disease is determined by the extent to which Cheju Islanders pollute their free environment with their own excreta. The widespread use of the pigsty-privy structures on the island greatly reduces the random dispersal of the tapeworm eggs in the free environment, and concentrates them in a toxic, walled compost pit where most eggs will perish unless eaten by the coprophagous pig. The structures are designed so that the potential of pork tapeworm infection becomes concentrated in the Cheju environment.

Park and Chyu (1963) surveyed 463 Cheju households (2,179 islanders), of which 93% kept privy-pigs. Ninety-six percent of these kept only one pig, and the rest, two pigs. Members of 52% of the households surveyed admitted to occasional consumption of raw pork. Sixteen percent of the islanders interviewed said that they were harboring tapeworms at the time of the survey. The subjects claiming tapeworm infection were predominantly male, and the highest infection rate was among those between the ages of 25 and 34 years of age. Raw pork eaters were the islanders having the highest tapeworm infection rates. There was no attempt to discover how many islanders suffered from cysticercosis (caused by accidental ingestion of the tapeworm egg from the environment). This study, however, concluded that approximately 20% of the Cheju Islanders harbored *Taenia* tapeworms. This high infection rate contrasted sharply with a 2% infection rate on the Korean peninsula.

In a subsequent and larger survey (1982), Kim Sung-ho discovered that 22% of the 1,020 rural islanders he surveyed harbored a tapeworm. Seventy percent of these islanders raised privy-pigs, and 30% of them were raw pork eaters. Kim's subsequent research effort (1984a) attempted to identify the cysticercus that infected the pigs, based on its typical morphology (shape). Kim explained his overall research procedure (1984b) as the examination of 61,420 hog carcasses that were slaughtered in Cheju's abattoirs over a 15-month period. Of these, 114 carcasses were infected with *Taenia solium* cysticerci, and all of those infected were determined to be privy-pigs. Kim also established that the pigs that were not infected were fed no human feces. In addition, older pigs, regardless of sex, had higher cysticercus infection rates. Thus, there is a persuasive logic in slaughtering a privy-pig on or near its first birthday. The pig's growth rate is greatest during its first year. It follows that energy input (pig feed) for caloric output (pork protein eaten) is most efficient when the pig is slaughtered at one year of age (Han, Kim, and Kang 1971).

Some of the data from the above surveys reflect islander attitudes about their controversial pig-keeping tradition. Asked to list the merits of their pigsty-privies, islanders interviewed by Park and Chyu (1968:167) indicated that the structures and system (1) contributed to the safe, clean, and convenient disposal of human excreta in the living environment; (2) reduced the feeding costs of raising pigs; (3) provided a safe and energy-saving method of composting manure essential to successful agricultural production; and (4) eliminated the odor of human feces in the living environment and reduced the excessive breeding of flies. These are, in the main, the same merits listed in a survey published by Kim Sung-ho (1982:70). The major disadvantage of the pigsty-privy system, according to the islanders, was the role it played in the transmission of parasite infections. They also registered some minor concern about the visual blight of the structures and other salient features of the system. Perhaps interviewees citing health concerns have been repeating the government's propaganda against privy-pigs.

The fact that a privy-pig must be trained while young to eat fresh human excreta (Park and Chyu 1963:166) would seem to confirm that Cheju islanders have a conscious and active, rather than a casual and passive, concern in maintaining this traditional waste-recycling scheme. They are aware of the importance of their privy-pigs as fertilizer factories, but cannot or do not articulate the possible health benefits of the system. However, Korean landscape epidemiologists have confirmed that some of the more dangerous and debilitating parasites infecting humans and pigs on the Korean mainland are significantly less threatening on Cheju Island, and some of these researchers attribute this difference to the positive attributes of the island's pigsty-privy system. For example, Kim Dang-chan (1970:305) examined 2,695 mainland Koreans for parasite infections and reported very heavy incidence of roundworms (70%), whipworms (68%), threadworms (23%), and hookworms (17%). Park and Chyu (1968:175) reported for Cheju that the incidence of roundworm infestation was only 47%; whipworms, 61%; threadworms, 4%; and hookworms, 8%. Their conclusion was that there was a low helminth, but high taeniasis prevalence on Cheju Island as compared with the Korean peninsula, and that the pigsty-privy system on Cheju acted "to lower the parasite infection level except for *Taenia*" (1968:161).

Conclusions

Although some aspects of the story that archaeologists tell about pigs in prehistory are rapidly changing, a consensus remains that pigs in prehistory were important as pork. Kim's establishing the role of pigs in the rise of political elites in Neolithic China is a good example of a consensus presumption about the primary role of pigs-as-pork in prehistory that is still highly speculative and open to challenge. Pigs-as-pork is a salubrious contemporary story that happens to promote economic growth, and is underpinned to some extent by latent Malthusian anxieties about the lack of animal protein as a limiting factor on human populations. Popular conversation may also extend into everyday archaeological discourse about the importance of pigs-as-pork in prehistoric subsistence. Thus Kim's speculation on the emergence of complex societies in prehistory presumes control and concentration of pigs-as-pork. The provocative ethnographic analogy from Cheju Island in South Korea suggests the importance of privy-pigs in subsistence for fertilizer production and disease-control, and not primarily for nutrition. It suggests, for example, that the evolution of pathogenic organisms may have posed a more serious threat to human survival in East Asian subsistence agricultural systems than protein shortages (Russell 1956:468).

Privy-pig systems may have emerged in prehistory as an efficient, productive and safe waste-recycling system that promoted socio-economic stability in subsistence. This raises an interesting question about the relative importance of the roles of pigs in stabilizing prehistoric subsistence systems: Would privy-pigs have contributed more to stabilization and durability of successful subsistence farming than pigs-as-pork? Kim's political thesis shifts scholarly attention and energy prematurely beyond the still-incomplete study of the complex role of pigs in subsistence agriculture. For example, egalitarian subsistence farmers' control of privy-pigs in a dispersed rather than concentrated manner, for the purposes of fertilizer production and disease control, could have nurtured the stability and gradual emergence of more complex societies. In this case privy-pigs served Neolithic Chinese as a source of stability in subsistence rather than "as funds of power for controlling exotic sumptuary items in order to achieve chiefly political hegemony" (S.O. Kim 1994:119).

Raish (1992) has studied differences in stability and the rise of complex societies in both the Old and New Worlds. She argues that subsistence-level farming economies with domesticated animals are more stable, measured in terms of duration, than those without these animals. Her studies suggest a shorter duration sequence for the pre-state farming period in the New World (e.g., Central America) when compared to the Old World (e.g., East Asia). She attributes this difference in duration to the contributions of medium and large domesticated animals to economic stability in the Old World. Pigs are among those animals present in the Old World, but absent in the pre-contact New World. Her study suggests that subsistence with a more stable economic component may undergo less rapid development of organizational complexity.

Kim's study on pigs in Neolithic Shandong relates to Raish's findings. Both Kim and Raish use cross-cultural approaches, but Kim argues also from ethnographic analogy and archaeological evidence the importance of pigs in Neolithic subsistence leading to the rise of a complex society. He also reasons that "small circular pits found in great numbers...functioned as structures designed for raising pigs" (1994:123). However, his story of increasing cultural complexity in Neolithic China links interaction between pigs as nutrition, and ritual, wealth, and political prestige. He also suggests that pigs were used as a medium of long-distance exchange for other kinds of prestige goods, which is his version of the pigs-go-to-market story. In his reply to his commentators, Kim writes: "The main thesis of my paper is that pigs in Neolithic China not only were important for human diet but also functioned as a means of social production. It seems that most commentators agree with me on this basic premise" (1994:138). Thus, Kim's thesis favors a consensus archaeological discussion that is overwhelmingly about pigs going to market (pigs-as-pork). Only Nelson offers an alternative story about pigs staying home (privy-pigs).

Did Neolithic Chinese have pigsty-privy systems? If so, did these systems nurture economic stability? Did they make a major contribution to the slow, steady growth of cultural complexity in Chinese agricultural civilization? Is this long-term contribution commemorated by the pigsty-privy figurines excavated from Han dynasty tombs? This paper has given close attention to the Cheju Island pigsty-privy system in order to provoke some response to these questions. The Cheju Island case study provides an ethnographic analogy that may increase archaeologists' understanding of the multiple roles of pigs in subsistence agricultural practices related to the development of cultural complexity in Neolithic China. Since the privy-pig structures (if not the pigs) are still abundant on Cheju Island, archaeologists might begin by examining their form and layout and exploring some comparisons with the Han grave models of pigsty-privies and with Kim's "small circular pits."

Were there pigs in prehistory? Archaeologists who

construct typologies and site maps from excavations most certainly believe so. Archaeologists who study the past in terms of the present, using ethnographic analogies are also quite convinced. However, critical archaeologists who examine archaeological conversations about the past cannot be so certain. To the extent that the conversation about pigs in prehistory is about pigs-as-pork rather than about pig life, it is a conversation that may no longer care to discern any difference between a life that is lived and a life worth living. The lives in question here are pigs', not humans', but examining the prevailing archaeological conversation on pigs-as-pork in prehistory draws attention to disturbing ethical trends and questions of humanity that are being shaped by contemporary social, economic and political forces—and not by forces in the past.

Note

1. During the early 1970s, I observed mostly small, lean black privy-pigs throughout Cheju Island. The traditional island agricultural system also utilized a local breed of miniature horses for plowing and transportation. I am suggesting that in the subsistence farming system of Cheju Island (as perhaps in Neolithic China, and until recently in North Korea and on the Ryukyu Islands), miniature breeds of domestic animals were appropriate low-technology solutions to problems of human survival. Perhaps these same breeds were once also used by Neolithic Chinese farmers; some examples of *Sus scrofa* of small size in East Asia include (feral?) black Chinese mountain pigs. These average 15 inches to 17 inches in height and 40 to 60 pounds. The small, lean black privy-pigs I have seen on Cheju Island fit the lower measurements in this description. Of course there is also the example of the Chinese pot-bellied miniature pig (now found only in a domesticated state).

References

Anderson, M. P. 1914. Forty Days in Quelpart Island. *Overland Monthly* (San Francisco) n.s. 63, 4(April): 392–401.

Anonymous. 1998. Editorial: The Battle over Hog Factories. *New York Times,* July 8.

Bray, F. 1984. Agriculture. *Science and Civilisation in China,* Vol. 6, Pt. 2 (Joseph Needham). Cambridge University Press, Cambridge.

Chaille-Long, C. (Col.). 1890. From Corea to Quelpaert Island: In the Footsteps of Kublai-Khan. *Bulletin of the American Geographical Society* 22(2): 218–266.

Cho, H.J. 1979. An Ethnographic Study of a Female Diver's Village in Korea. Ph.D. dissertation, Anthropology, University of California, Los Angeles.

Glacken, C.J. 1955. *The Great Loochoo: A Study of Okinawan Village Life.* University of California Press, Berkeley.

Groves, C. 1981. *Ancestors for the Pigs: Taxonomy and Phylogeny of the Genus* Sus. Technical Bulletin No. 3, Department of Prehistory, Research School of Pacific Studies, Australian National University. Canberra.

Han, S.W., H.K. Kim, T.S. Kang. 1971. A Basic Study on the Improvement of Swine Feeding on Cheju Island. *Cheju National University Faculty Research Journal* 3:285–299.

Hegner, R., F.M. Root, D.L. Augustine, C.G. Huff. 1938. *Parasitology.* D. Appleton-Century Company, New York.

Heiser, C.B., Jr. 1973. *Seeds to Civilization: The Story of Man's Food.* W. H. Freeman, San Francisco.

Holmes, J.C. 1976. Host Selection and Its Consequences. In *Ecological Aspects of Parasitology*, ed. C.R. Kennedy, pp. 27–33. North-Holland Publishing Company, Amsterdam.

Hsu, C.Y. 1980. *Han Agriculture.* Jack Dull (ed.). University of Washington Press, Seattle.

Izumi, S. 1971. *Cheju Do.* Tokyo University, Tokyo.

Kim, D.C. 1970. Studies in the Control of Ascariasis and Hookworm Infection by Periodic Mass Treatment. *Report of the National Institute of Health, Republic of Korea*, Vol. 7, pp. 281–306. National Institute of Health, Ministry of Health and Social Affairs, Seoul, Korea.

Kim, S.H. 1977. Survey of Taeniasis on Cheju Island. *Cheju National University Faculty Research Journal: Natural Sciences* 9:83–87.

——— 1982. Survey of Taeniasis Infection and Eating Habits of Raw Pork on Cheju Island. *Cheju National University Faculty Research Journal: Natural Sciences* 14:65–70.

——— 1984a. Morphology of Cysticercus Cellulosae Rostellar Hooks and Epidermis Found on Cheju Island. *Cheju National University Faculty Research Journal: Natural Sciences* 17:71–85.

——— 1984b. Relationship Between Cysticercus Cellulosae Infection and Feeding Conditions of Swine on Cheju Island. *Cheju National University Faculty Research Journal: Natural Sciences* 17:103–111.

Kim, S.O. 1994. Burials, Pigs, and Political Prestige in Neolithic China. *Current Anthropology* 35(2): 119–133.

Laufer, B. 1962. *Chinese Pottery of the Han Dynasty.* 2d ed. Charles E. Tuttle Co., Rutland, VT.

Lee, Y.K. 1994. Comments on: "Burials, Pigs, and Political Prestige in Neolithic China." *Current Anthropology* 35(2): 133–135.

Nelson, S.M. 1994. Comments on: "Burials, Pigs, and Political Prestige in Neolithic China." *Current Anthropology* 35(2): 135–136.

Nemeth, D.J. 1985. Cheju Island's Pigsty-Privies: The Architecture of Sincerity. *Landscape* 28(3): 15–21.

——— 1987. *The Architecture of Ideology: Neo-Confucian Imprinting on Cheju Island, Korea*. University of California Press, Berkeley.

——— 1989. Commentary: A Study of the Interactions of Human, Pig, and the Human Pork Tapeworm. *Anthrozoos* 3(1): 4–13.

——— 1995. Discussion and Criticism: On Pigs in Subsistence Agriculture. *Current Anthropology* 36(2): 292–293.

Oelhal, R.C. 1978. *Organic Agriculture: Economic and Ecological Comparisons with Conventional Methods*. John Wiley and Sons, New York.

Oliver, William L.R., Colin P. Groves, Roger C. Cox, and Raleigh A. Blouch. 1993. Origins of Domestication and the Pig Culture. In *Pigs, Peccaries, and Hippos*, ed. William L.R. Oliver, pp. 171–179. International Union for Conservation of Nature and Natural Resources, Gland, Switzerland.

Park C.B., and I. Chyu. 1963. A Socio-Epidemiological Study of the Swine Pen Human Latrine System Practiced on Cheju Island. *Journal of the Catholic Medical College* 7:161–186.

Pitts, F., W.P. Lebra., and W.P. Suttles. 1955. *Post-War Okinawa*. Pacific Science Board, National Research Council, Washington, DC.

Quilter, J. 1994. Comments on: "Burials, Pigs, and Political Prestige in Neolithic China." *Current Anthropology* 35(2): 137.

Raish, C. 1992. *Domestic Animals and Stability in Pre-State Farming Societies*. Tempus Reparatum, Oxford.

Rappaport, R. 1967. *Pigs for the Ancestors: Ritual in the Ecology of a New Guinea People*. Yale University Press, New Haven.

Rosman, A. 1994. Comments on: "Burials, Pigs, and Political Prestige in Neolithic China." *Current Anthropology* 35(2): 137–138.

Russell, R.J. 1956. Environmental Changes Through Forces Independent of Man. In *Man's Role in Changing the Face of the Earth*, ed. W. L. Thomas et al., pp. 453–470. University of Chicago Press, Chicago.

Schmidt, G.D., and L.S. Roberts. 1981. *Foundations of Parasitology*. 2d ed. The C.V. Mosby Company, St. Louis, MO.

Simoons, F.J. 1961. *Eat Not This Flesh: Food Avoidances in the Old World*. University of Wisconsin Press, Madison.

Spencer, J.E. 1954. *Asia, East by South*. John Wiley & Sons, New York.

Temple, R. 1986. *The Genius of China: 3,000 Years of Science, Discovery, and Invention*. Simon and Schuster, New York.

T'angso [History of the T'ang Dynasty]. Compiled A.D. 945 by Liu Hsu.

Tyler, P.E. 1997. Plague Ravages Taiwan Pigs and Many Blame China. *New York Times*, April 19 (Friday), p. 5.

Wilford, J.N. 1994. First Settlers Domesticated Pigs Before Crops. *New York Times*, May 31 (Tuesday), p. B5, B9.

Wilson, R.F. 1996. Pig. *The Encyclopedia Americana: International Edition*. Vol. 22, pp. 86–89. Grolier, Danbury, CT.

AN ETHNOGRAPHIC VIEW OF THE PIG IN SELECTED TRADITIONAL SOUTHEAST ASIAN SOCIETIES

P. Bion Griffin

Department of Anthropology, University of Hawai'i, Honolulu, HI 96822

In the ethnographic present of Southeast Asia, the pig stands out, Chief of a diverse tribe of wild and domestic food animals. For hunters, the wild pig is *the* desired prey. Of course, nearly any mammal, reptile, amphibian, bird, or aquatic animal that can be secured will most likely be shot, and eaten or exchanged. For farmers, the domesticated hog is the mainstay of animal husbandry efforts, and the largest animal easy to breed and grow.[1] Cultivators—tribal swiddeners, fixed field farmers, peasants—favor *Sus scrofa* and may complement its production with domestic fowl, cattle, and dogs. Hunters usually limit their domestic beasts to dogs, and dogs are critical in most scenarios of wild pig acquisition. Two themes dominate my views of pigs in this paper. First, we must understand pigs in the context of different tribal economic strategies, and second, we must consider the suite of animals exploited in addition to, and in a few cases in lieu of, the domestic pig *Sus scrofa*, and the most important wild (not feral) pig, *Sus barbatus*, the "bearded pig" (Fig. 1).

Insular, peninsular, and mainland Southeast Asia house a great variety of peoples and cultures, as well as an extremely diverse flora and fauna. The genus *Homo* has been in the region nearly two million years if recently discovered fossils and archaeometric dates are to be trusted. Foraging societies reached the most isolated portions—eastern Indonesia and the Philippines—at least by the late Pleistocene. Farming societies have existed in favorable habitats during much of the Holocene. Many aspects of the subsistence strategies of present-day foragers and farmers are instructive to the archaeologist. Since the people of Southeast Asia have been interacting

Fig. 1. The bearded pig, *Sus barbatus*. The scalp has been removed.

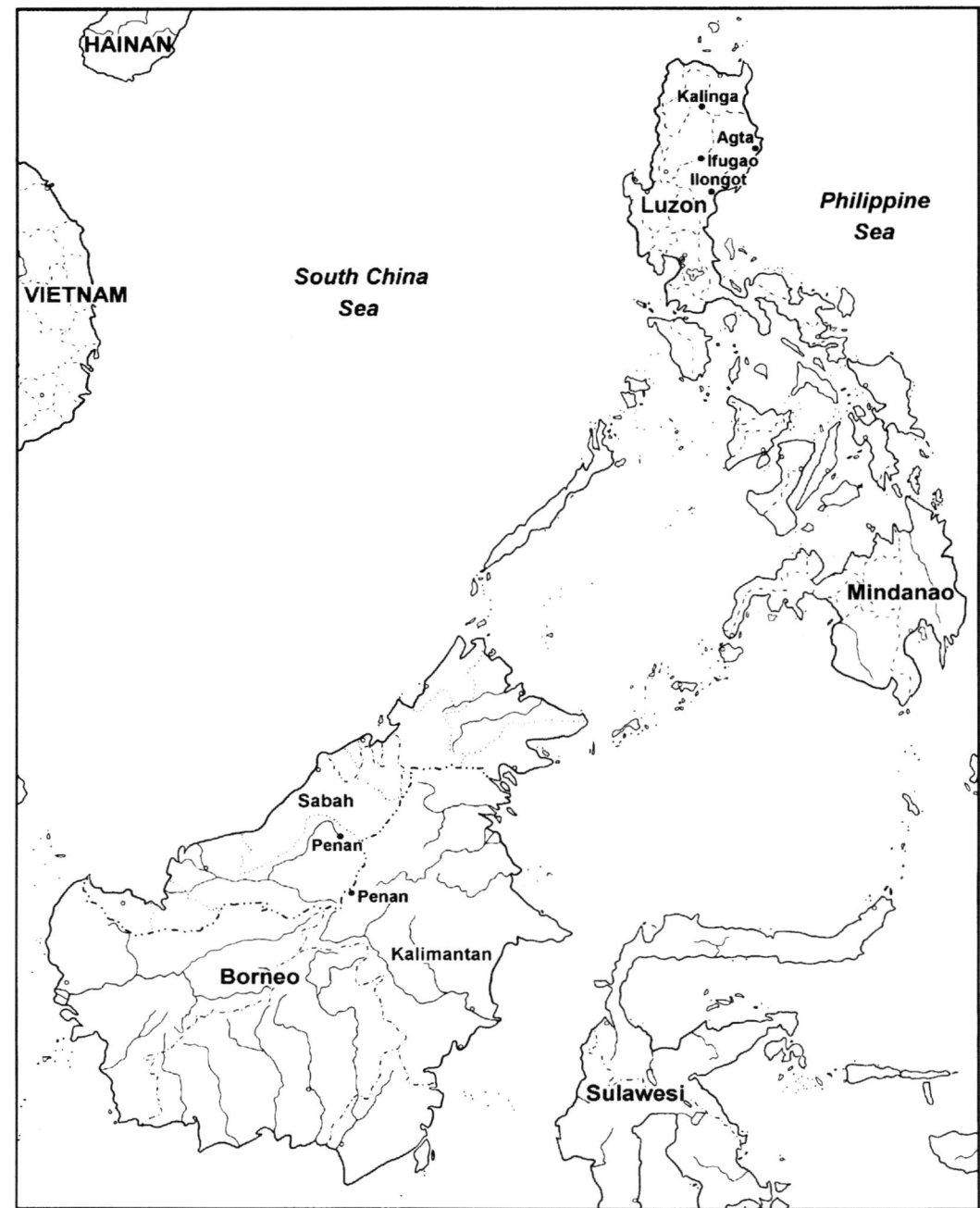

Fig. 2. Map of Southeast Asia and location of groups mentioned.

with pigs throughout much of this time, an examination of variations in this dependency is useful.

The Role of Pigs in Foraging Societies

Pigs and people begin their relationship among hunters-gatherers, or foragers. The Agta, a tribal people of northeastern Luzon in the Philippines (Fig. 2), among whom my family and I have lived off and on since 1972, represent one end of the ethnographic spectrum (M.B. Griffin 1996; P.B. Griffin 1989; Griffin and Estioko-Griffin 1985. For a bibliography, see Headland and Griffin 1997). Agta pride themselves as hunters of wild pig (*Sus barbatus*, not feral *Sus scrofa*; Mudar 1986, 1985) and deer, as fishers, and as deeply knowledgeable forest collectors. Pigs are by far the most desired kill and are the most economically and nutritionally important animal; their procurement influences much of the organization of Agta society.

The wild pig is important because it provides much of the needed protein and fat that sustain the Agta and some of their non-Agta neighbors; it also supplements the limited carbohydrates in their diet. Deer simply do not store body fat in the fashion of pigs, which may have a good 5 cm of fat over much of their bodies during the rainy season. The only other sources of animal fat are pythons, which may have sheaths of fat inside their body cavities, and riverine fish. Pythons are few and difficult to secure. Fish, however, are a rich source of fat during the dry season, when pigs are usually lean. Riverine fish are usually lacking in fat during the cold rainy season and are seldom procured since the rivers are cold, flooded, and silted, making spear fishing difficult. During the dry season, rivers become clear, warm, and plentiful in fish and eel. Fish, in fact, seem as essential for nutrient supply and as exchange goods as are pigs. Were fish unavailable, the well-being of the Agta or other foragers would be markedly diminished. Fish and wild pigs are truly complementary.

Pig fat and fatty fish are doubly important for Agta since the natural or wild supply of plant-based carbohydrates is limited. The wild yam hypothesis aside (the idea that tropical forests lack plant carbohydrates [Headland 1987]), the wild roots, seeds, and nuts of Luzon are not a rich and ready resource for foragers. Sweet potato and cassava,[2] plus maize and rice, have for decades provided most of the starch for Agta consumption, although most of these are gained through trade with farmers. Certainly, at various times and places wild roots have been important; even during our fieldwork, collecting wild plant food was necessary in times of scarcity, but usually acquisition of food from farmers and other non-Agta was possible.

For the Agta, wild pigs are therefore both a source of much protein and, seasonally, nearly all fat, and are the basis of much of the Agta's ability to secure by trade basic plant foods, metals for tools, and the consumer goods desired. Of course, in the remote past and in times of crisis such as during World War II, little except metal for knives was needed beyond plant foods. Not even cloth was obtained, bark cloth sufficing well into this century. Informants characterize this period as one of little deprivation, since game and fish populations were adequate and foragers could use these in exchange for plant foods.

Much of the pattern of subsistence and settlement was and is built around a combination of access to wild pig and to trade partners. Agta live in extended family residential clusters, usually along rivers or streams, perhaps at the coast, and some distance from trade partners. The most remote "mountain" Agta may travel a day or more to reach an exchange point; most Agta, however, must travel only one, two, or three hours. Increasingly, trade partners camp at the Agta homes, hoping to acquire a catch before others do, or before the Agta can eat much. They usually have arrived with rice, so that the Agta are immediately in their debt.

Hunting Strategies

Agta hunting practices involve a variety of tactics and strategies. In the rainy season, men may hunt singly or in small parties. Stalking, ambushing alongside fruit trees, and jacking (spotting a prey animal's eyes with a flashlight beam) are options. During this season the forest is quiet, the pigs fat and less wary, and stalking is relatively easy, if uncomfortable because of the cold rains. Dogs, unenthusiastic about forest travel with its mud, cold, and discomfort, are nearly useless at these times. They must usually be carried across streams and cannot ford rivers. They follow pig and deer spoor with difficulty. During the dry season, however, pigs are lean, are alert for forest noises such as the tread of humans, and are quick to take flight. Dogs are especially useful during these periods, since they travel quickly and quietly, picking up pig scent readily. Dogs are trained to drive startled game to hunters waiting in ambush along trails. Both pig and deer are frequent quarry, although monkeys may occasionally be trapped in an isolated tree by really skillful dogs.

Dogs are a valuable resource for capturing pig and deer, albeit an expensive one. They enable hunting to be a year-round activity for the Agta. Dogs enable Agta to *specialize* in pig and deer hunting and to de-emphasize swidden cultivation. Agta have food materials to trade every month because they have a hunting technology that is effective through all seasons. Dogs do have to be fed food that humans could eat, such as scraps from animal carcasses and leftover rice or roots. This is a strain on Agta well-being, since seldom, at least since we have been observing, is their own nutrition adequate. In addition, dog populations are not self-sustaining. Dogs are occasionally killed by pigs, and due to malnutrition, they are often sickly and die from disease; animals must therefore be purchased from farmers who can feed their dogs with grains. However, dogs are considered mandatory by the Agta we know since they function as early warning systems, creating an uproar when raiders from other groups or tribes are prowling about. And all Philippine tribal groups we have lived among or queried insist that raiding is, or was, a way of life. Pigs and dogs are together an essential component of the lives of hunters, swiddeners, and farmers.

Fig. 3. Agta home with wild piglet captive inside.

Consumption and Distribution of Hunted Meat

The work of Navin Rai (1990) on the western slopes of the Sierra Madre and our own work on the eastern slopes and coast reveal a similar picture of wild pig capture and utilization. In the remoter areas of habitation and hunting, a family group, usually with three to five adult male hunters, sometimes with female hunters, secures about 25% of its caloric intake from hunted meat and about 50% from food acquired by trade. Fishing varies markedly by season, but often fish is consumed daily. Women and children may catch small fish for meals when hunters are gone simply by spending a few minutes at the river or stream adjacent to the camp, spearing bottom fish, shrimp, and eel. Among the Disabungan Agta, in a sample of 64 days of observation and 53 days of hunting, Rai recorded 48 hunting trips and 21 pigs killed, but only 9 deer and 5 monkeys. These Agta traded 217 kg of pig meat, but consumed 305 kg. Deer was traded in greater proportions: 171 kg exchanged versus 56 kg consumed. These Agta found retention of a certain amount of pig fat mandatory.

I note that in the preceding and following references to meat and fat, these are lumped together, as they are in Agta-arranged trade packets. Agta may select especially fatty pieces of meat, or pieces including fat, bone and skin for either exchange or consumption. Usually a selection of lean and fat is included in any "share," although Agta tend to keep the less attractive (less fleshy and fatty) cuts for their own consumption. Agta may consume the entire pig, excepting the digestive tract (usually taken by farmers) and the lungs (given to dogs). Fat is not separated from meat except in an occasional cooking process where the lard is rendered for consumption with boiled plant foods. The skull and vertebrae are cooked and almost entirely picked of edible materials by Agta alone. Simply put, for both Agta and non-Agta, a portion consisting of a fatty piece of meat and some marrow-rich bone is ideal. The usual meal of meat consists of a starch and boiled pieces of meat, skin with fat and lean attached, and bones chopped with a machete, to which meat adheres.

During our stay in 1975 among Ihaya Agta, a group related to the Disabungan people, a higher portion of pig meat and fat was consumed by the Agta themselves, yet slabs of belly fat were set aside for farmer trade partners. In late 1980 through early 1982, among a northerly unrelated group, the Nanadukan Agta, the local pig population had fallen due to over-hunting and an adverse typhoon season, and most pigs killed were traded for rice and consumer goods. Interestingly, the Nanadukan Agta were more successful cultivators than the Disabungan and had access to superb fishing resources. They had also developed a greater demand for consumer goods. Production of a portion of their own plant foods and a willingness to sacrifice some animal fat did permit their way of life to continue, but they sorely felt the lack of wild pigs. They even discussed, but rejected, the possibility of raising an occasional captured wild pig for resale. (The practice is occasionally attempted among Agta [Fig. 3], but ends in an early sale of an immature pig.)

Other Patterns of Subsistence

Archaeologists, especially in Southeast Asia, might hypothesize that as soon as foragers took up a serious amount of plant cultivation, pig husbandry would begin. I suggest that such is not the case. First, the tropical forests of the region house wild pigs in great abundance.[3] No advantage is gained by raising pigs, and some disadvantages may be argued. Pig husbandry seems to develop as access to wild pig meat decreases, either because habitat is lost or because hunters who provide the resource themselves disappear. In fact, in Palanan, Isabela province, one may today observe an increase in dependence on domestic pigs as wild pigs become scarce, as Agta foragers abandon their traditional way of life, and as farmers seek reliable sources of meat and fat.

Combining Hunting and Swiddening

The complexities of integrating pig hunting and very small scale farming are minimal when pressure on land is unknown and when few non-hunters are nearby. Agta informants tell us that decades ago they often cleared small swiddens at scattered locations throughout their hunting ranges, returning to them for harvest of whatever had survived and grown. These plots were then beyond the grasp of lowlanders or upland tribal swiddeners, although the crops suffered the depredations of animals. Today, the Agta do not enjoy these conditions. Lowland farmers tend to seize swidden plots cleared by Agta, who have little recourse. Their response in recent years has been to decrease mobility as they opt for larger and more productive gardens and fields. This increased sedentism inhibits travel to productive hunting areas and leads to overexploitation of game within walking distance. Even today, Agta crops are so small, when they exist at all, that little impact is felt in scheduling hunting and fishing.

The Ilongot, neighbors of some Agta of northeastern Luzon, exemplify tribal swiddeners who hunt pig and deer, yet who depend on swidden cultivation of upland rice, sweet potatoes, cassava, and dry taro, plus assorted vegetables and fruits. From studies by several scholars, especially William Jones, a pioneer ethnographer of 1908, and the Rosaldos working in the late 1960s, we have a good view of Ilongot subsistence and society (Jones 1907–09; R. Rosaldo 1981, 1980). By the 1990s their way of life had radically changed, but earlier in this century and until the time of martial law in 1972 they often hunted pigs and deer daily, using bows and arrows like Agta, but also spears, unlike Agta. They were not dependent on farmers for plant foods, although they did seek beads, metals, cloth, and consumer goods through exchange. By using dogs to hunt pigs in the dry season, they competed with Agta on an equal footing, and they often used spears to dispatch their prey, thereby lessening the mortality rate of their dogs. As among the Agta, dogs were used to secure wild pigs, as well as to warn of approaching raiders. By emphasizing bow and arrow hunting in the rainy season, the Ilongot maintained a high intake of fat, supplementing their stores of rice and sweet potato.

Ilongot are especially interesting because they seldom raised pigs in spite of being swiddeners. Douglas Yen (pers. comm. 1996) observed that Ilongots kept an abundance of hunting dogs. He suggests that the fierceness of the dogs would have endangered any domesticated pigs living around the houses. I believe the dogs were also sentries, since the Ilongot were, several years ago, even more prone to raids than were the Agta. The security of the homesteads would have been seriously compromised if the number of dogs had been reduced in order to make pig-keeping viable.

Chickens were raised, but were consumed for ritual reasons. Ilongot did not build rice irrigation systems, but moved their homes to new swiddens as the old were exhausted. This pattern, coupled with the natural abundance of the bearded pig, did not favor swine husbandry. One anthropological puzzle remains unsolved concerning Ilongot subsistence and Agta exchange patterns. Two dialect groups of Agta are neighbors of the Ilongot, and interacted with them in various ways. William Jones (1907–09) reported that Agta were enemies of Ilongot, with the latter usually afraid of the deadly forest-dwelling Agta. A model of Agta as dependent on exchange, however, raises questions about their relationships with Ilongot. More recently, my own queries during encounters with these Agta (whose language I speak poorly) brought incomplete answers depicting cooperation among the two groups. My suspicion, based on historical materials collected by Thomas Headland, is that Agta traded with more distant, fixed-field farmers and tended their own modest swiddens, replicating the Ilongot way of life on a simpler scale. I see the foraging Agta and the foraging/swiddening Ilongot as little different, with dependence on the wild pig as central to the viability of both. As competitors for that most valuable of resources, they likely placed themselves apart, usually maintained adversarial relations, and found trade partners among other groups.

An even more striking example of the providence of pigs for hunters is found among the Penan of Borneo. Peter Brosius (1991), working in Sarawak, Malaysia, and Rajindra Puri (1997), living in Kalimantan, Indonesian Borneo, report Penan hunters as potentially independent of farmers. Penan, living in groups somewhat larger than Agta and most Ilongots, hunt primarily the bearded pig. The Borneo bearded pig is a variety larger than that of the Philippines (Mudar 1997), and it is found in large migrating herds. In the Sierra Madre of Luzon, pigs move in small family clusters, sometimes a mother and her young, or several sisters and their young, or so the Agta say. They do not range widely, moving over a fixed territory from the coast to well up the mountain slopes. Agta report that the pigs range over distances of perhaps 10 km at best. In Borneo, herds of hundreds of pigs migrate seasonally over many kilometers, crossing major rivers, then retrace their routes. Penan slaughter the migrating pigs by the dozens, and they hunt the year around, using dogs and spears. Their kill per hunt ratio is not uniformly high, since seasonality and pig behavior influence hunting success. The bow and arrow complex is unknown. The Penan choice of spears may be associated with (1) the greater abundance of pigs at any one moment, (2) the opportunities to spear pigs while they are fording rivers, and (3) the greater safety for hunting dogs, who may easily run in front of cast arrows. Spears are a more effective technology when considering Borneo pigs; the bow and arrow seem more appropriate in the Philippine hunter-gatherer context. Quoting Puri (1997:157-158):

> The bearded pig (*Sus barbatus*) was by far the most abundant mammal species caught by residents of Long Peliran during the study (N=707) contributing 79.8 percent of all catches and 91 percent of all edible meat yielded... Assuming a mean population of 100 residents, the bearded pig provided potentially an average of 366 grams of meat per person per day during the 21 month study.

> In reality of course, the Penan families consume a majority of these catches, while the Kenyah families [neighboring swidden farmers] depended on fish and domestic chickens to a much greater extent. Both groups experienced periods of absolutely no meat from the bearded pig as well as periods of great abundance. For instance, there was not a single pig caught between the middle of February and the 6th of May 1991. Periods of abundance were primarily due to huge daily harvests taken during two large pig migrations that passed through the area.

These figures would stagger Agta and Ilongot hunters, as would the size of the pigs—often twice that

Fig. 4. An Ifugao pig house and pen. Photograph by Wilhelm G. Solheim II

Fig. 5. Domestic pigs, Dangtalan, Kalinga.

of the Philippine variety. While the Agta seldom kill an adult boar weighing 50 kg, Penan kill pigs approaching 100 kg. In addition, the Penan adaptation is striking when compared to Philippine swiddeners. There is seemingly little need to exchange with Kenyah, at least for plant food. Penan harvest the extensive stands of wild sago palm (*Arenga undulatifolia*), which provide a starch staple of nearly limitless quantities, plus numerous other forest plant foods. The Penan certainly find no advantage in pig husbandry, nor will they as long as the forests are provident in bearded pig and sago palm. Only as these resources are lost due to deforestation and expanding farming populations, or when more cash is needed to gain consumer goods, may the Penan lose their relationship with wild pigs.

Fixed-Field Farming

The subsistence strategies of the Ifugao of the Philippines represent a development away from pig hunting and swiddening. They are tribal fixed-field farmers famous for their rice terraces, many varieties of rice, complex social organization, and rich traditions. Certainly the Ifugao of 1996 are not as representative of tribal fixed-field farmers as they once were—like all Filipinos, they are lawyers, nurses, engineers, politicians, and migrants around the world. Yet, looking at the ethnographic past, we may see their use of pigs as quite different from Ilongots, Agta, and Penan.

Traditional Ifugao live in the central Cordillera of Luzon, occupying hamlets and farmsteads scattered near valley bottoms and up mountain slopes. A variety of anthropologists have described these people, including such famous figures as Roy Barton (1922) and Harold Conklin (1980). My own observations are based on the literature, including Jenks (1902) among the Bontoc, on conversations with ethnographer Maria Stanyukovich, and a few weeks' residence among Ifugao migrants in the Sierra Madre.

Ifugao may be at times successful hunters of feral pigs, *Sus scrofa*, which are found in stands of forest throughout the Cordillera. Ifugao tend to hunt with dogs and spears, not the bow and arrow. As a result, they hunt in drier weather. Since their abundant rice harvest is brought in before the heavy rains, less scarcity is found during the ensuing months. Further, the wild pig as a source of sustenance is largely replaced by the domestic hog. Pig pens are sometimes a feature of Ifugao homes that we have visited (Fig. 4), and in places where crop security is not an issue, pigs range and forage nearby the residences. As breeding stock of Euro-American sources are introduced, pigs are increasingly penned, being less vigorous than the native variety.

Among the Ifugao, pigs are especially important in ritual feasting and in marking status. Funerals, which among Ilongots would entail wild pig and domesticated chicken, among Ifugao may favor home-grown pigs in considerable numbers.

Not surprisingly, the lesson of the Ifugao is that with sedentary farming homesteads, fixed or permanent field systems, and reliance on one's own cultivated plant foods comes a diminished involvement with hunting. Two processes develop for these farmers. First, animal husbandry increases, with emphasis usually on the domesticated pig, and second, continued access to wild meats increasingly favors exchange with culturally different forest-dwelling hunters. The hunters, like the Agta but unlike the Penan, need the exchange relationship since they cannot produce enough of their own crops. Indeed, the farmers actively discourage hunters' farming efforts in order to retain the hunters' foraging services and to keep potentially arable land to themselves. But the farmers, in order to maintain their part of the bargain and gain wild foods, must plant surpluses for exchange. One way of diminishing the dependence on hunters is to raise pigs. This is always a compromise, since pigs consume some food humans would eat and they demand care and attention.

Still, in Southeast Asia among tribal farmers, the trend has been to exploit the domesticated pig more and more. This variation among subsistence emphases is found among groups other than Agta, Ilongot, Penan, and Ifugao. The Hanunoo of Mindoro, Philippines (Conklin 1957), are somewhat similar to Ilongot, although far less fierce in reputation. Hanunoo swiddeners raise a few pigs, which compete with humans for food, including sweet potato. In addition, swiddens are fenced to keep out both domestic and wild pigs. The latter are trapped when possible. Hanunoo domestic pig husbandry falls somewhere between Ilongot and Ifugao in pattern, if not in trajectory of development.

I'wak, who live somewhat south of the Ifugao homeland in Luzon, may represent a further trend toward "the Ifugao model." I'wak are fixed-field farmers, although perhaps unique in their relative de-emphasis on rice production; instead they favor taro cultivation (Peralta 1982). Domestic hogs are kept in most households, usually living under the family's domicile at night (Peralta 1982:46). In addition to cultivated foods, pigs are fed gathered plant foods such as bamboo shoots and palm hearts. As with lowland farmers, sedentary tribal farmers process banana stalks as pig fodder. Tiruray of Mindanao (Schlegel 1979) differ in that they regularly feed pigs largely cultivated crops, including papaya, but do not consume the meat, selling the animals alive or butchered. The Kalinga, a sedentary rice farming society immediately north of the Ifugao, depend little on hunted pigs, although variation is found depending on the abundance of wild populations. As noted by Takaki in her detailed ethnography (1977:151):

> Pigs freely roam in and around a settlement during the day [Fig. 5], foraging and, on the way, cleaning up garbage and like matters from settlement areas. In addition to self-feeding in the course of their daily rounds, they are fed twice a day, in the morning and the evening, with the pig food known as *qanal* which is the mixture of garbage, left-over

Fig. 6. Agta trophies of wild pig mandibles.

food, additional vegetable foods gathered or harvested, and chipped rice and rice bran all cooked together with a quantity of water... Responding to the call of their keepers announcing their evening meal, pigs return home and sleep under the raised floor of the house. In the morning, they stay around the settlement until fed and then depart again for a day of foraging.

Conclusions

We learn from several different Southeast Asian peoples the overwhelming importance of pigs, be they wild or raised. Pig flesh and fat is a vital component of the diet in these tropical lands. As hunting and hunted animals disappear, pig husbandry and intensification of cultivation increase. Pig husbandry is necessarily tied to enhanced food production, since pigs within settlements are not self-sufficient. As is well documented among farming societies in New Guinea, domesticated pigs compete for human food, necessitate human labor expenditure in crop raising and in fodder processing, and demand care in handling and in health maintenance (Dwyer 1990; Rappaport 1984; Rubel and Rosman 1978). Moreover, among farmers such as Ifugao and Kalinga, pigs are seldom consumed simply as meat on the daily table, but are nearly always first channeled through ceremonial rites, filling ideological and social functions before nutritional and gastronomic uses.

Foragers such as Agta and swiddeners such as Ilongot tend to consume more pig meat, albeit wild, hunted pig. They eat considerably less plant foods, excepting perhaps a major starch, neglecting fresh or cooked greens and vegetables. The ceremonial use of pigs is concomitantly minimal. Ilongots favor chickens in ritual; Agta at best offer morsels to spirit beings either after a game kill or at a meal, when a drink or a piece of food is announced as available to nearby ghosts.

Foragers' and swiddeners' campsites and homesteads reflect the lack of pig husbandry, evidencing no temporary or long-term pens within, under, or adjacent to dwellings. The ground surface is disturbed in patterns peculiar to agents other than pigs, who are notorious rooters. Dogs often push into the warm but dead embers of campfires in order to gain warmth, and people may dig pits, but pigs create soil churning and wallows second only to the mud pools of water buffalo.

Especially worth considering are people's relationships with pigs and dogs. Foragers and swiddeners typically do not consume dog flesh. First, dogs are too valuable to eat. They are essential for hunting in the humid tropics. Second, dogs are essential to all tribal peoples in Southeast Asia because they are so effective in warning of approaching strangers. As noted above, the problems associated with dogs include their lessened usefulness for hunting in the rainy season and their high mortality. Puppies are difficult to rear since the mother is often lacking in milk, due to her terrible diet. Disease strikes dogs at all ages and especially in times of nutritional stress (which among the Agta is constant). The introduc-

Fig. 7. Ifugao house with trophies consisting of pig mandibles and a water buffalo skull.
Photograph by Wilhelm G. Solheim II

tion of domestic pigs or captive baby wild pigs is problematic, since dogs attempt to kill them. Initial attempts to adopt pig husbandry include close watching of pigs by children and constructing small pens adjacent to houses, as well as beating dogs into learning that pigs in camp are off-limits.

In Southeast Asia, trophying of mandibles or skulls of wild and domestic animals is a custom often observed. The custom is useful to both ethnographers and archaeologists in their quests to understand pigs and people. For example, Agta and Ilongot hunters kept pig mandibles on roof rafters or on rattan straps beside the houses (Fig. 6), while Ifugao are famous for keeping mandibles and crania of assorted mammals as trophies (Fig. 7). These customs provide one of several material bases for looking at variation in the relationships of pigs to settlements, social organizations, and subsistence systems. The pig is ubiquitous in prehistoric, historic, and traditional Southeast Asian societies, and has proven itself, along with the dog, central to viable ways of life throughout the region.

Notes

1. I exclude animal growers who are Muslim from any comments concerning pigs.

2. Sweet potato, cassava (manioc), and maize were introduced to the Philippines after A.D. 1521 by the Spanish colonizers.

3. In the 1990s much forest has been lost to logging and clearance for farming. In earlier decades, virgin forest was well populated by the bearded pig.

References

Barton, Roy F. 1922. *Ifugao Economics.* University of California Publications in American Archaeology and Ethnology 15(5): 385–446.

Brosius, J. Peter. 1991. Foraging in Tropical Rain Forests: The Case of the Penan of Sarawak, East Malaysia (Borneo). *Human Ecology* 19(2): 123–150.

Conklin, Harold C. 1957. *Hanunoo Agriculture: A Report on an Integral System of Shifting Cultivation in the Philippines.* Food and Agriculture Organization of the United Nations, Rome. Reprinted by Elliot's Books, Northford, CT.

——— 1980. *Ethnographic Atlas of Ifugao.* Yale University Press, New Haven.

Dwyer, Peter D. 1990. *The Pigs That Ate the Garden: A Human Ecology from Papua New Guinea.* University of Michigan Press, Ann Arbor.

Griffin, M.B. 1996. The Cultural Identity of Foragers and the Agta of Palanan, Isabela, the Philippines. *Anthropos* 91:111–123.

Griffin, P. B. 1989. Hunting, Farming and Sedentism in a Rain Forest Foraging Society. In *Farmers as Hunters: The Implications for Sedentism*, ed. Susan Kent, pp. 60–70. Cambridge University Press, Cambridge.

Griffin, P.B., and Agnes Estioko-Griffin (eds.). 1985. *The Agta of Northeastern Luzon: Recent Studies.* San Carlos University Press, Cebu City.

Headland, Thomas N. 1987. The Wild Yam Question: How Well Could Independent Hunter-Gatherers Live in a Tropical Rainforest Ecosystem? *Human Ecology* 15:463–491.

Headland, Thomas N., and P. Bion Griffin (comps.). 1997. A Bibliography of the Agta Negritos of Eastern Luzon, the Philippines. SIL Electronic Working Papers 1997-004. (http://www.sil.org/silwp/1997/004/silwp1997-04.html). Summer Institute of Linguistics, Dallas.

Jenks, Albert E. 1905. *Bontoc Igorot.* Bureau of Printing, Manila.

Jones, William. 1907–1909. The Diary of William Jones: 1907–1909. Robert F. Cummings Philippine Expedition. Unpublished field notes, Chicago Natural History Museum, Chicago.

Mudar, Karen M. 1985. Bearded Pigs and Beardless Men: Predator-Prey Relationships Between Pigs and Agta in Northeastern Luzon, Philippines. In *The Agta of Northeastern Luzon: Recent Studies*, ed. P. Bion Griffin and Agnes Estioko-Griffin, pp. 69–84. San Carlos University Press, Cebu City.

——— 1986. *A Morphometeric Analysis of the Five Subspecies of Sus Barbatus, the Bearded Pig.* M.A. thesis, Dept. of Zoology, Michigan State University, East Lansing.

——— 1997. Patterns of Animal Utilization in the Holocene of the Philippines: A Comparison of Faunal Samples from Four Archaeological Sites. *Asian Perspectives* 36(1): 67–105.

Peralta, Jesus T. 1982. *I'wak Alternative Strategies for Subsistence. A Micro-Economic Study. The I'wak of Boyasyas, Nueva Vizcaya, Philippines.* Anthropological Papers No. 11. National Museum, Manila.

Puri, Rajindra K. 1997. Hunting Knowledge of the Penan Benalui of East Kalimantan, Indonesia. Ph.D. dissertation, Department of Anthropology, University of Hawai'i, Mänoa.

Rai, Navin K. 1990. *Living in a Lean-to: Philippine Negrito Foragers in Transition.* Museum of Anthropology, University of Michigan, Ann Arbor.

Rappaport, Roy A. 1984. *Pigs for the Ancestors.* Enl. ed. Yale University Press, New Haven.

Rosaldo, Renato. 1980. *Ilongot Headhunting 1883–1974: A Study in Society and History.* Stanford University Press, Stanford.

——— 1981. The Social Relations of Ilongot Subsistence. In *Adaptive Strategies and Change in Philippine Swidden-based Societies,* ed. Harold Olofson, pp. 29–42. Forest Research Institute, College, Laguna, Philippines.

Rubel, Paula G., and Abraham Rosman. 1978. *Your Own Pigs You May Not Eat: A Comparative Study of New Guinea Societies.* University of Chicago Press, Chicago.

Schlegel, Stuart. 1979. *Tiruray Subsistence: From Shifting Cultivation to Plow Agriculture.* Ateneo de Manila Press, Quezon City.

Takaki, Michiko. 1977. Aspects of Exchange in a Kalinga Society, Northern Luzon. Ph.D. dissertation, Department of Anthropology, Yale University, New Haven.

EVALUATION OF MOLAR SIZE AS A BASIS FOR DISTINGUISHING WILD BOAR FROM DOMESTIC SWINE: EMPLOYING THE PRESENT TO DECIPHER THE PAST

John J. Mayer

Westinghouse Savannah River Company, P.O. Box 616, Aiken, SC 29802

James M. Novak and I. Lehr Brisbin, Jr.

Savannah River Ecology Laboratory, P.O. Drawer E, Aiken, SC 29802

Introduction

The domestication of swine (*Sus scrofa* L.) and other animal species was a keystone cultural achievement of early human populations, and identifying the time(s) and place(s) of its occurrence has been an important goal of archaeological investigations. The ability to accurately identify wild ancestors from truly domestic forms in associated faunal remains is often based on the presence of derived or altered morphological characters in comparison to the wild ancestor (Bökönyi 1969; Zeuner 1963). Skull characteristics, which have been widely recognized by taxonomists as one of the best means of classifying vertebrates (Lowe and Gardiner 1976), have been among the most important traits used to document the domestication process (Bökönyi 1969; Clutton-Brock 1981; Zeuner 1963). Unfortunately, cranial and mandibular material recovered from archaeological sites is often insufficient to allow either quantitative or qualitative comparisons.

Unlike the skull, teeth are frequently preserved intact in prehistoric sites due to their compact and dense structure. Since dentition is almost invariably affected by proportional size changes in the skull, measurement of tooth size has long represented an alternative technique for identifying cranial size changes resulting from domestication (Bökönyi 1974; Zeuner 1963).

Domestic swine are all descended from a single species, the Eurasian wild boar. Domestication of *S. scrofa* has been reported as having occurred independently in a number of sites ranging from Europe to the Far East (Clutton-Brock 1981; Epstein 1971; Flannery 1961; Keller 1902; Kowalski 1976; Kuşatman 1992; Pira 1909; Staffe 1922; Zeuner 1963). Thus, a number of different subspecies of Eurasian wild boar (Fig. 1) would have been collectively ancestral to modern-day domestic swine. The earliest known domestication of this species is estimated to have taken place in the region encompassing the Middle East and eastern Europe between 6,000 and 8,000 B.C. (Clutton-Brock 1981; Epstein 1971; Herre and Rohrs 1990; Kuşatman 1992; Zeuner 1963).

As in several other species, the transition from ancestral wild boar to derived domestic forms of swine entails a shortening of the rostral region of the cranium and associated changes in the mandible (Bökönyi 1974; Clutton-Brock 1981; Epstein 1971; Kelm 1938; Kowalski 1976; Mayer and Brisbin 1991; Pira 1909; Rutimeyer 1862; Stampfli 1983; Zeuner 1963). Such brachycephalic alterations have also resulted in the subsequent shortening of the molariform dentition (Flannery 1961; Stampfli 1983). Because teeth preserve well, the lengths of the second and third molars have received a fair amount of attention in distinguishing wild vs. domestic swine. This widely recognized difference has led to the use of crown length in these molars for identifying the time course for domestication in swine, with greater molar lengths being considered to represent wild forms (Amschler 1939; Bökönyi 1974; Flannery 1961, 1983; Higham 1968; Lawrence

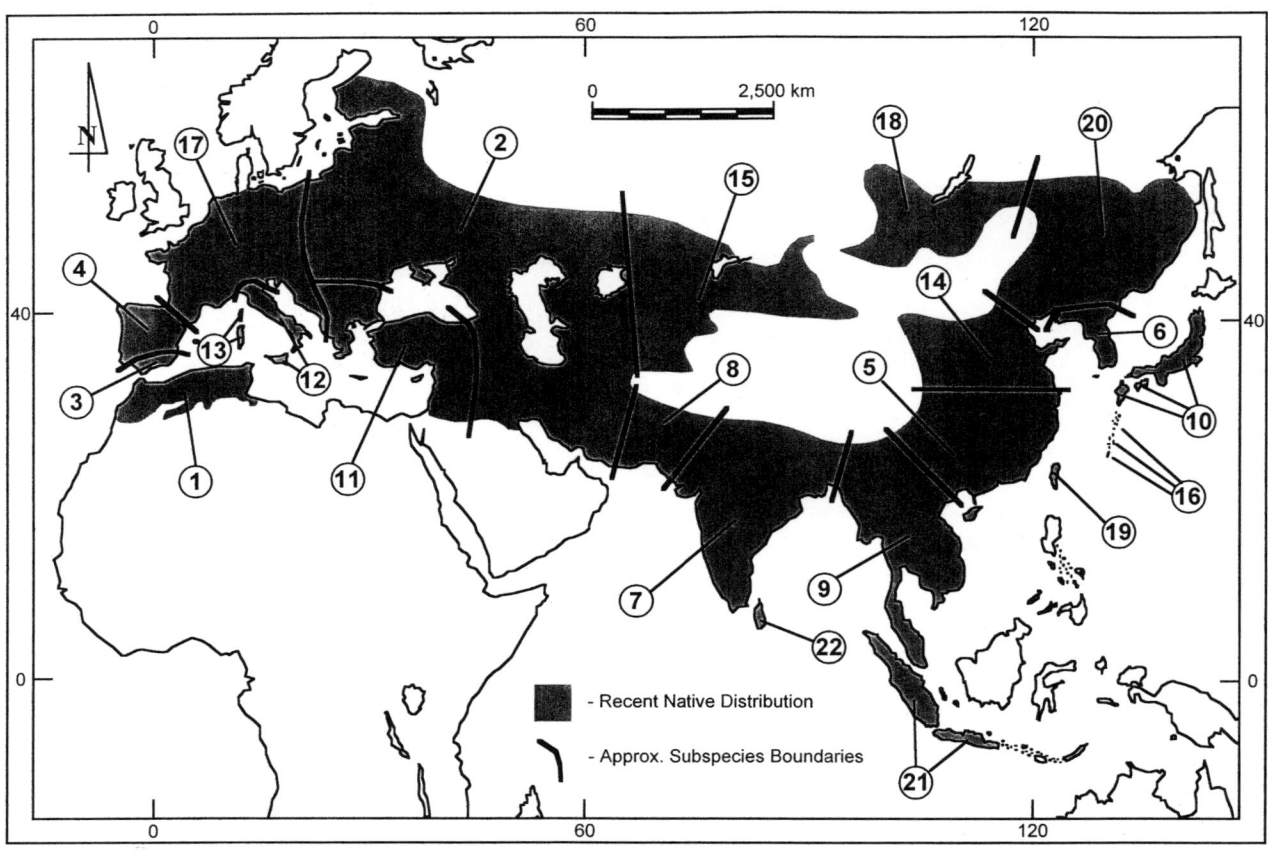

Fig. 1. Present-day distribution (shaded area) of Eurasian wild boar (*Sus scrofa* spp.) with approximate subspecies boundaries. The subspecies are as follows: (1) *S. s. algira;* (2) *S. s. attila;* (3) *S. s. baeticus;* (4) *S. s. castilianus;* (5) *S. s. chirodontus;* (6) *S. s. coreanus;* (7) *S. s. cristatus;* (8) *S. s. davidi;* (9) *S. s. jubatus;* (10) *S. s. leucomystax;* (11) *S. s. lybicus;* (12) *S. s. majori;* (13) *S. s. meridionalis;* (14) *S. s. moupinensis;* (15) *S. s. nigripes;* (16) *S. s. riukiuanus;* (17) *S. s. scrofa;* (18) *S. s. sibiricus;* (19) *S. s. taivanus;* (20) *S. s. ussuricus;* (21) *S. s. vittatus;* and (22) *S. s. zeylonensis*. Data modified from Mayer and Brisbin 1991 and Oliver, Brisbin, and Takahashi 1993.

1980; Reed 1969; Stampfli 1983; Stein 1989). In spite of the widespread use of this technique, the size variation of these teeth within and between the various types of *S. scrofa* has not been sufficiently studied to validate its continued application.

The purpose of this paper is to evaluate the use of second and third molar length and width as a diagnostic basis for identifying archaeological specimens of domestic swine. This study compares variation observed in recent *S. scrofa* material and uses it to identify and evaluate analogous changes in the morphological transition from the wild ancestor to derived domestic forms of swine. To further examine the man-made progression from wild ancestor to derived domestic, a known crossbred form between these two types was used to determine if very early derived morphological types (i.e., transitional forms) could also be distinguished from the wild ancestor. Within and among these three types, the variations attributable to sex and age were analyzed as potentially confounding parameters. In addition to univariate differences in molar size, the relationship of molar allometry (i.e., width vs. length) was also analyzed to determine if significant differences exist among the three types of swine. Finally, if molar size differences are found between wild and domestic swine, these data could be used to develop a more statistically sound method for distinguishing between types.

Assessment Approach

The typical approach taken by researchers in applying the aforementioned technique is to use specimens of regionally indigenous Eurasian wild boar to determine a minimum size for the second and third molars (Flannery 1961, 1983; Higham 1968; Stampfli 1983). These size limits are then used to establish a wild-domestic size threshold.

This commonly used approach is based on the assumption that there has been no translocation of non-native wild individuals or stocks of different sizes into the regions being studied. The potential for such translocation events is a potentially confounding possibility which raises questions about the use of only indigenous specimens to establish a wild-domestic molar size threshold. The sudden appearance of physically smaller swine might not necessarily always be the result of the importation or development of derived domestic forms. The importation of a smaller captive or tamed Eurasian wild boar cannot be totally discounted. Similar sudden size changes among domestic stocks, for example, have previously been attributed to just such introductions of outside sources (Zeuner 1963). In general, the size variation among the subspecies of Eurasian wild boar is gradual. However, abrupt size differences between adjacent subspecies do exist. A few immediately adjacent mainland subspecies exhibit marked size differences (e.g., between *S. s. scrofa* and *S. s. attila*, and between *S. s. castilianus* and *S. s. baeticus*) (Groves 1981; Kuşatman 1992; Mayer and Brisbin 1991). Size contrasts between mainland and nearby island subspecies can also be significant. Potential translocations of smaller adjacent subspecies would throw doubt on the exclusive use of a larger indigenous subspecies in the determination of a wild vs. domestic threshold based upon molar size.

It should be noted that at this time we are not aware of any case where such translocations of non-native subspecies have occurred or impacted any conclusions in the analysis of a local archaeological site. However, since the long-distance transportation of early domestic swine is widely thought to have taken place (Clutton-Brock 1981; Zeuner 1963), would it not also have been possible for immature or juvenile captive wild boars to have been carried or transported along ancient trade routes to be sold in distant lands? It therefore may be prudent to employ a broader representation of the variation seen among the different subspecies of *S. scrofa*.

In addition to potential effects of translocation events, sexual dimorphism in size exhibited by Eurasian wild boar creates a further confounding aspect in determining a valid wild-domestic threshold. Size dimorphism between male and female Eurasian wild boars has been well documented (e.g., Briedermann 1970; Harrison 1968; Hell and Paule 1983; Heptner, Nasimovic, and Bannikov 1966; Koslo 1975; Mayer and Brisbin 1991, 1993; Payne and Bull 1988; Romic 1975) and is significant in adults (Hell and Paule 1983; Mayer and Brisbin 1991). The possibility thus exists for the larger males to be identified as Eurasian wild boars and the smaller females as domestic swine on the basis of molar size alone. A similar size difference between the sexes has been a complicating factor in distinguishing wild from domestic cattle (Grigson 1969, 1982a, 1982b).

Finally, teeth appearing during the intermediate stages of the dental eruption pattern within swine have the potential to decrease in crown length due to abrasion from adjacent teeth (Payne and Bull 1988). Such age-related factors would have the potential to affect the crown length of the second molar in going from immature to adult specimens and could compromise the validity of any wild-domestic size threshold.

These aforementioned factors (i.e., geographic, sexual dimorphism, and age-related variability) require a broader look at the crown length and width variation exhibited by the molars of the different types of swine. If sufficient molar size differentiation indicative of wild vs. domestic status were still found to exist, a method that would be more robust to these factors could be produced. On the other hand, if the overlap of size between wild vs. domestic forms is large relative to these factors, this would seriously question the use of molar size as a basis for distinguishing wild from domestic swine.

In the present study, data from known extant forms are used to define the variability found in these teeth and then to evaluate the validity of using the second and third molar mensural size and width allometry as the basis for identifying wild from domestic swine. Recent specimens of known Eurasian wild boars are currently available for morphological analysis. However, the molar size and shape of either very early domestic swine or of the transitional forms between the wild ancestor and prehistoric domestic forms of swine are not well documented.

Due to the paucity of intact and readily identifiable specimens, the morphological appearance of an early domestic swine phenotype is difficult to determine with certainty. Recent/modern-day domestics (i.e., from the late 1880s through the present day), although admittedly distinct from Eurasian wild boars, are also almost certainly different from prehistoric domestic swine. In comparison to the four recent major types of *S. scrofa* (i.e., Eurasian wild boar, recent/modern domestic swine, feral swine, and wild boar x feral swine hybrids), archaeological cranial material of domestic swine (e.g., as illustrated in Bökönyi 1974; Keller 1902; and Pira 1909) qualitatively most closely resembles feral swine. Feral swine are defined as wild *S. scrofa* whose ancestry is solely from domestic swine (Mayer and Brisbin 1991). In canonical variates analyses of crania, specimens of prehistoric domestic swine fell within the recent feral swine target group (Fig. 2). Many feral swine populations have been wild-living for three to five centuries, and a few for more than 1000 years (Mayer and Brisbin 1995). Feral swine populations on

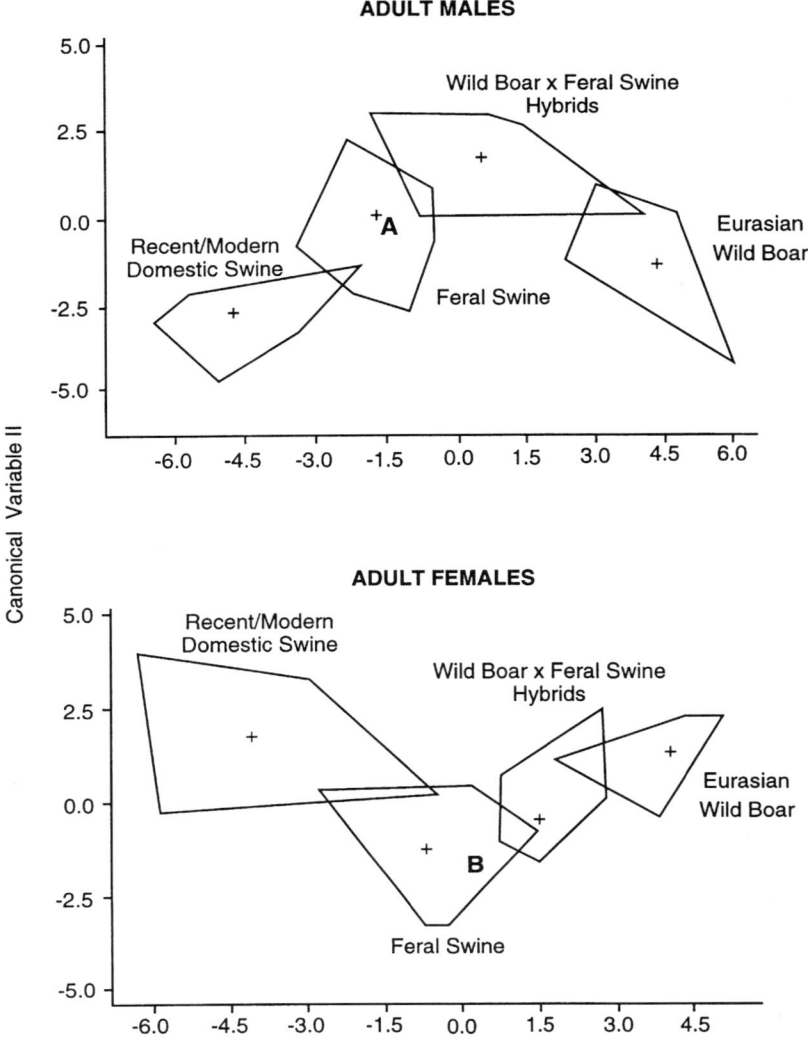

Fig. 2. First two canonical variables for adult male (top) and female (bottom) crania of *Sus scrofa* comparing the four general types with two domestic specimens (male = A; female = B) from an archaeological site in Uppsala, Sweden, dating back to the Middle Ages (measurements taken from Pira 1909). Canonical variable plots based on Mayer and Brisbin 1991.

Ossabaw Island off the southeastern coast of the United States and on the Andaman Islands in the Bay of Bengal are two such examples (Brisbin 1989, 1990; Mayer and Brisbin 1995; Oliver and Brisbin 1993). Some of these feral populations have in fact been found to resemble early or former domestic breeds which no longer exist under husbandry conditions. A few such feral populations have even been found to represent the last remnants of long-gone domestic breeds (Mayer and Brisbin 1995; Van Vuren and Hedrick 1989). Therefore, because of the close morphological resemblance of the skull, we propose that recent or modern-day feral swine are probably the best present-day analog or surrogate for early prehistoric domestic swine.

We further propose that present-day wild boar x feral swine hybrids are the most appropriate surrogate for the transitional stage between Eurasian wild boar and early domestic swine. This type of *S. scrofa* represents a morphological middle ground between the two extremes of ancestral wild boar and derived feral/domestic stock. Hybrids, possessing a mixture of both wild boar and feral/domestic characters, would also resemble a potential evolutionary middle ground occupied by the transitional stage described above.

This study will use the phenotypic variation of recent individuals of wild boar, hybrids, and feral swine as sequential stages in a surrogate model to characterize the morphological transition from wild to early domestic individuals respectively (Fig. 3). In this model, the feral swine and hybrid specimens would be used as morphological surrogates for the early domestic swine and the wild/domestic transitional form, respectively.

Materials and Methods

A total of 937 recent specimens (198 Eurasian wild boars, 212 wild boar x feral swine hybrids, and 527 feral swine) were examined during this study. The sample of Eurasian wild boars consisted of recent museum specimens collected from various locations throughout the native distribution of the wild species in Europe, North Africa, and Asia. These specimens included representatives of 21 of the 22 normally recognized geographic subspecies of Eurasian wild boar (Fig. 1). The one subspecies not represented in this sample was *S. s. riukiuanus*. The feral swine and wild boar x feral swine hybrid samples included both museum specimens and animals recently collected in the field. Most of the feral specimens came from mainland and island populations in the United States; however, museum specimens representing feral populations from the Andaman Islands, Australia, Belize, Costa Rica, Galapagos Islands, Gardener Island, Mariana Islands, Mexico, Nicobar Islands, New Zealand, and Pemba Island were also included. Despite their varied origins and scattered distributions, the general phenotype of the feral swine skull is consistently uniform and identifiable as such in comparison to that of the other recent major types of *S. scrofa* (Mayer and Brisbin 1991, 1993). All of the wild boar x feral swine hybrid samples came from populations in the United States. Specimens within each of the three groups consisted of varying combinations of crania and mandibles. In addition to the aforementioned samples, 79 recent/modern-day domestic swine were included for comparative purposes with the Eurasian wild boar and the two morphological surrogate samples. A listing of the specimens obtained from existing collections is provided in Mayer and Brisbin 1991. Those specimens not contained in that reference were recently collected in the field, and are currently contained in the senior author's personal holdings.

Based either on the known past history of the source population or analysis of the individual specimen's cranial morphology (Mayer and Brisbin 1991, 1993), each of these specimens was categorized into one of the three types of swine as follows: Eurasian wild boar, wild boar x feral swine hybrids, and feral swine. They were also identified as to both sex and age class. Sex was determined by the data provided on the museum specimen tag, examination of fresh specimens collected in the field, or the morphology of the canines for museum/pick-up specimens of unknown sex (Mayer and Brisbin 1988). Age class categories included yearling, subadult, and adult, and were based on erupted dental patterns as described in Mayer and Brisbin 1991. Lacking either of the molars being studied, animals younger than the yearling age class were not included.

Linear measurements were taken from both the upper and lower second and third molars present in the specimens, with teeth measured on the right side of the specimen where possible. Measurements were made with 150-mm dial calipers to the nearest 0.1 mm. Ten

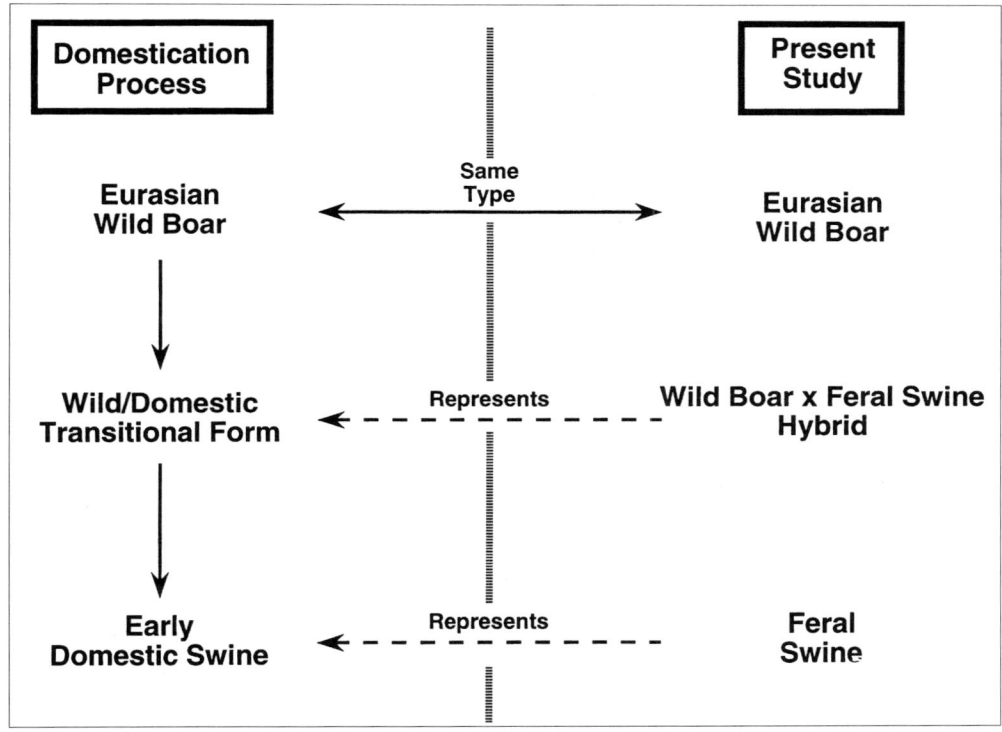

Fig. 3. Schematic representation of the approach used in the present study to classify groups of present-day *Sus scrofa* as morphological surrogates for earlier stages in the initial domestication of the species. The recent Eurasian wild boar sample is considered to be equivalent to the archaeological representatives of this same type of *Sus scrofa*.

Table 1. Sizes (in mm) of second and third molars of the three types of *Sus scrofa*. Sex and age class data were combined within each type

Type of Swine	Molar	Measurement	N	Mean	Observed Range	SE
Eurasian Wild Boar	Upper 2nd	Length	190	22.6	17.5–28.8	0.13
		Width	190	19.2	13.7–26.5	0.10
	Upper 3rd	Length	156	35.2	24.0–50.0	0.27
		1st Width	168	21.6	15.0–30.0	0.14
		2nd Width	160	18.9	12.6–27.0	0.13
	Lower 2nd	Length	191	21.8	16.0–27.9	0.13
		Width	191	16.0	11.9–20.3	0.09
	Lower 3rd	Length	154	39.0	27.0–53.0	0.32
		1st Width	170	18.0	13.0–24.5	0.12
		2nd Width	166	17.4	13.2–22.5	0.11
Wild Boar x Feral Swine Hybrid	Upper 2nd	Length	182	21.2	15.4–25.9	0.13
		Width	182	16.8	14.4–19.8	0.10
	Upper 3rd	Length	78	31.5	25.2–40.0	0.38
		1st Width	109	19.2	12.8–23.5	0.17
		2nd Width	86	16.3	13.9–19.3	0.17
	Lower 2nd	Length	154	20.7	17.2–24.1	0.15
		Width	154	14.2	11.6–19.2	0.10
	Lower 3rd	Length	60	33.5	24.6–41.4	0.51
		1st Width	93	16.0	13.3–18.7	0.16
		2nd Width	85	15.5	12.6–17.4	0.15
Feral Swine	Upper 2nd	Length	361	20.3	15.5–24.5	0.09
		Width	361	16.5	13.2–20.5	0.07
	Upper 3rd	Length	177	30.0	23.5–38.0	0.25
		1st Width	231	18.5	15.0–22.0	0.1
		2nd Width	201	16.1	12.8–19.0	0.11
	Lower 2nd	Length	475	19.8	12.3–27.7	0.08
		Width	475	13.8	11.2–18.8	0.06
	Lower 3rd	Length	198	32.0	24.2–40.3	0.28
		1st Width	299	15.3	12.2–20.2	0.09
		2nd Width	272	15.2	11.3–19.2	0.08

measurements (five on the upper and five on the lower tooth rows) were taken as follows: second molar length—the greatest length of the crown of the second molar; second molar width—the greatest width of the crown across the posterior cusp row of the second molar; third molar length—the greatest length of the crown of the third molar; width of first cusp row of third molar—the greatest width of the crown across the first cusp row of the third molar; and, width of the second cusp row of third molar—the greatest width of the crown across the second cusp row of the third molar. These measurements were taken consistent with the methods described in Driesch 1976 and Mayer and Brisbin 1991.

All statistical analyses were performed using the Statistical Analysis System (SAS) version 6.12 (SAS Institute Inc. 1989). All variables were analyzed for conformation to a normal distribution using a Shapiro Wilk test in Proc Univariate and normal probability plots. Analyses of variance were performed using the SAS Mixed procedure which allows both fixed and random effects in the models. All effects in these models were considered fixed, but the procedure provides statistics useful for model selection to allow discrimination of models with different terms and different numbers of terms. Analyses of covariance models were first fit with all relevant interaction terms of covariates (heterogeneity of

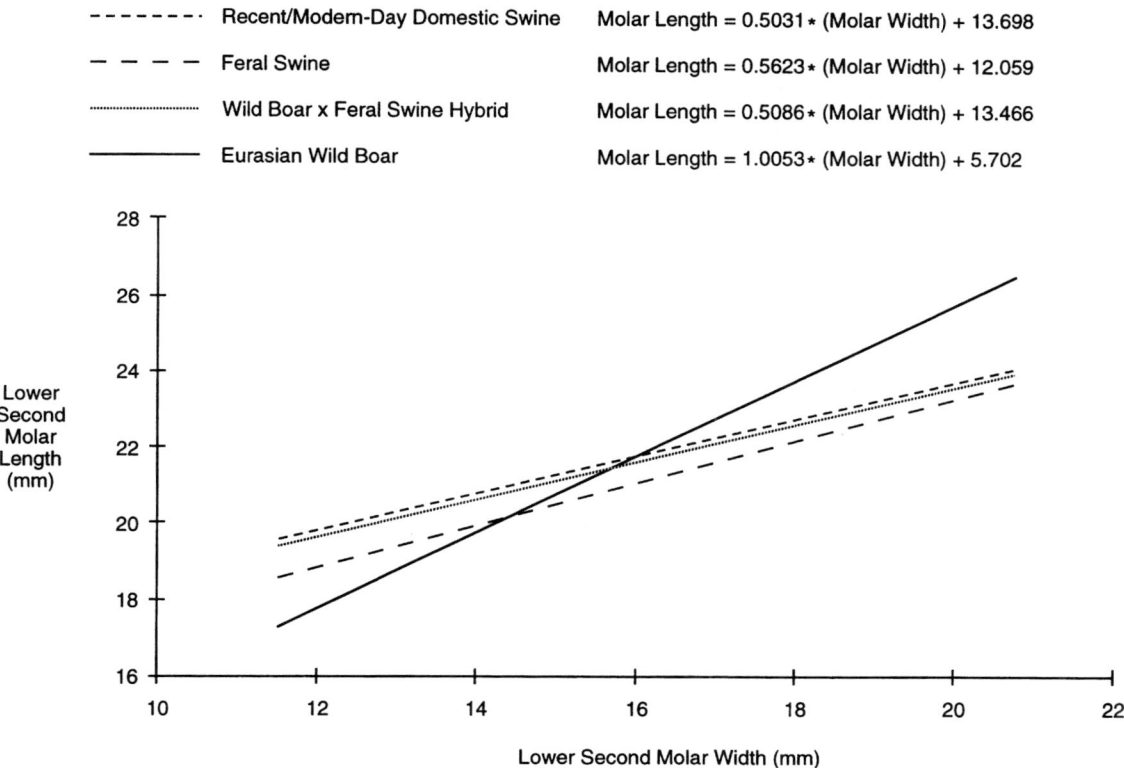

Fig. 4. Linear plots of lower second molar length and width (in mm) among the three types of swine used in the present study. Recent or modern-day domestic swine are included for comparison with the three types.

variance models). If the interactions were not significant, then the models were fit with only the main effects of the covariates included. Statistical significance was accepted at P<0.05, with acceptance criteria modified for multiple tests using a sequential Bonferroni procedure as required.

The wild-domestic thresholds for the various measurements were developed using the lower limits of the 95% confidence intervals for Eurasian wild boar, and the upper limits of the 95% confidence intervals for hybrids and feral swine. The use of this metric instead of the minimum and maximum limits of the observed range (e.g., Flannery 1961, 1983; Higham 1968; Stampfli 1983) provides a more statistically sound threshold. The use of range limits (i.e., minimum and maximum values) as identification thresholds entails the application of extreme observations to define differences between sample groups. The range overlap created by such extremes can obscure the valid separation of a measurement exhibited by the majority of specimens being analyzed.

Results

Summaries of the ten molar measurements for the three types of swine are presented in Table 1. The univariate differences among the three types (i.e., with sex and age classes combined) were significant for all of the molar lengths and widths. In each measurement, the wild boars were the largest, decreasing in size to the hybrids, and then followed by feral swine as the smallest. Therefore, the surrogate model was consistent within the molar size gradient among the three types of swine. In light of these significant differences, it should also be noted that range overlap by extreme observations did occur among the three types for all of the parameters measured (Table 1).

The overall differences in molar size between the sexes with the types combined were significant for all ten measurements. Males were consistently larger in all of these intersex size comparisons. However, within each type, although males averaged and ranged larger than females, the differences between the sexes were not found to be significant ($F=0.57$, $d.d.f.=714$, $p=0.56$). Thus, size differences due to sexual dimorphism would not represent a significant confounding aspect in determining the wild-domestic threshold.

Because of the initial absence and then eruption and presence of the third molar immediately posterior to the second molar (in going from yearling to subadult and then adult), one would expect that interproximal abrasion of the posterior face of the second molar crown would result in a decrease in crown length along this age class gradient. Second molar crown lengths were found to follow this expected pattern of decreasing size in successively older age classes within each type except for the Eurasian wild boar sample. This latter pattern was due to a seeming increase in second molar crown length in the subadult Eurasian wild boar sample. However, upon investigation, only the larger wild boar subspecies (e.g., *S. s. attila*) were represented among the specimens comprising this subadult sample. Thus, this apparent increase was likely an artifact of this particular data set. In general, however, like sexually dimorphic differences, age-related differences in molar size were not sufficient to obscure the overall difference between the types.

The allometric relationship of all molar widths and lengths was found to be significant among the three types of swine. The length-width relationships were consistent among the three types for each of the molars except in the lower second molar ($F=11.77$, $d.d.f.=814$, $p=0.0001$). The Analysis of Covariance of that tooth showed that the allometric relationship did not differ between the feral swine and hybrids, but that both of those were significantly different from the Eurasian wild boar (Fig. 4). This would indicate that molar allometry as well as size of the second lower molar could be used to distinguish wild ancestral vs. prehistoric domestic swine.

Based on the 95% confidence intervals, the wild and domestic thresholds for the various molar measurements are provided in Table 2. The upper prehistoric domestic threshold was defined as the upper limit of the 95% confidence interval for the hybrids. Although close in some parameters (e.g., lower second molar length), no overlap occurred in any of the measurements between the upper limit of the domestic and the lower limit of the wild boar 95% confidence intervals. Overall, the third molar length had the broadest gap between thresholds, and the second molar length the narrowest. For both of the third molars, the gap breadth was greatest for the crown length, followed by the second cusp row width, and then by the first cusp row width. A reverse pattern was observed in the second molars, with the cusp row widths having the broadest gaps, followed by the crown lengths.

The combined percentage of specimens incorrectly classified using the 95% confidence interval limits were also calculated for the three samples of known swine. The lowest percentage of incorrect identifications was for the upper second molar width (only 4.4%), followed by the second cusp row width of the upper third molar (5.6%). In spite of the fact that it had the broadest threshold gap, the lower third molar length had the next lowest percentage (7.3%) of incorrect identifications. The highest percentage of incorrect assignments of specimens was for the lower second molar length (22.8%), with upper second molar length (15.7%) as the next highest. The percentage of unknowns (i.e., specimens falling within the threshold between wild and domestic) was highest for the second cusp row width of the upper third molar (32.4%) and lowest for the lower second molar length (7.0%). Overall, if a measurement had a low percentage incorrectly classified, it tended to have a high percentage of unknowns among the three types of swine.

Discussion

The present study confirms the value of second and third molar size as a basis for differentiating wild boar from early domestic swine. However, the determination of valid thresholds for these molar lengths and widths needs to be predicated on the knowledge of the variation in these dental measurements exhibited in both wild ancestral and derived domestic forms of *S. scrofa*. Based on the results of the present study, the few critics of the validity of

Table 2. Listing of the maximum domestic and minimum wild size thresholds for each length and width measurement (in mm) of the upper and lower second and third molars of *Sus scrofa*

Molar	Crown Measurement	Maximum[a] Domestic Swine Threshold	Minimum[b] Wild Boar Threshold
Upper Second	Length	21.4	22.4
	Width	17.1	19.0
Upper Third	Length	32.2	34.6
	1st Width	19.6	21.4
	2nd Width	16.6	18.7
Lower Second	Length	21.0	21.5
	Width	14.4	15.8
Lower Third	Length	34.5	38.3
	1st Width	16.3	17.7
	2nd Width	15.7	17.2

[a] Based on the upper limit of the 95% confidence interval for hybrids
[b] Based on the lower limit of the 95% confidence interval for wild boar

these size differences in wild vs. domestic swine (e.g., Bolomey 1973; Chaplin 1969; Teichert 1969) would not appear to be completely justified.

Within the comparisons undertaken in the present study, Eurasian wild boar are a known morphological entity which can be carefully examined and defined on the basis of extant free-living populations worldwide. The current distribution of Eurasian wild boars ranges from the Iberian Peninsula to the Maritime Territory of Siberia. The observed size variation among the different subspecies of Eurasian wild boar is notable (Groves 1981; Kuşatman 1992; Mayer and Brisbin 1991). In fact, this size variation has been widely used to describe and distinguish the various geographic races of this species. A loose clinal situation appears to exist, with the physical body size (e.g., head-body length, shoulder height, snout length, hind foot length, etc.) of wild boar increasing somewhat to the north, and more significantly to the east.

The largest described subspecies include *S. s. attila* and *S. s. ussuricus*. The smaller subspecies are mostly represented by insular forms, including *S. s. meridionalis*, *S. s. taivanus*, *S. s. riukiuanus*, and specific Southeast Asian island populations of *S. s. vittatus*. This observation is consistent with the phenomenon of insular dwarfing documented in a number of other ungulate species occupying both mainland areas and islands (Case 1978; Foster 1964). It should also be noted, however, that at least some of the insular subspecies of wild boar (i.e., *S. s. leucomystax* and *S. s. zeylonensis*) do not seem to exhibit dwarfing effects in restricted insular habitats. Conversely, some continental or mainland subspecies are of relatively small size. These would include *S. s. baeticus* and *S. s. majori*. Several theories have been advanced to explain the species-wide variation in body size of Eurasian wild boar. The most common hypotheses center around a post-glacial intermixing of previously isolated larger northern and smaller southern forms (Ammon 1938). Habitat also appears to be a factor, with animals found in mesic habitats being larger than those found in xeric areas (Epstein 1971; Spitz, Valet, and Brisbin 1998). A similar size differentiation reportedly also occurs between populations found in mountains vs. plains habitats (Epstein 1971). Using a representation of most of the variation seen among the different subspecies, the present study provides thresholds which would have a broader application regionally than any of the previous studies that used molar size differences based mostly upon local subspecies.

With the size variation found in Eurasian wild boar, there would be some instances in which using the threshold values provided in Table 2 would be inappropriate for identifying domestic swine. Such circumstances include studies which encompass areas inhabited by those subspecies or populations of wild boar occupying the lower end of the physical size spectrum. Examples of this include *S. s. baeticus*, *S. s. meridionalis*, *S. s. taivanus*, *S. s. riukiuanus*, and specific island populations of *S. s. vittatus* (Fig. 1). The second and third molars of these wild boar are equivalent to or smaller than the 95% confidence intervals of both surrogates used in this study and samples of known prehistoric domestics. Thus, molar size could not be used validly to identify the presence of prehistoric domestic swine in lands inhabited by these wild boar subspecies.

Analysis of size variation in Eurasian wild boar skeletal or dental material from an archaeological setting is further complicated by the larger size of prehistoric specimens as compared to recent specimens from the same locations. The physical size of wild boar was determined to have decreased during the postglacial periods (Ammon 1938; Bökönyi 1974; Epstein 1971; Herre 1949; Kurten 1968; Kuşatman 1992; Stampfli 1983). Moreover, this decline in size has been noted to continue into present times, with series of specimens from the same locations generally appearing to become smaller from the 1800s through the 1900s (Heptner, Nasimovic, and Bannikov 1966; Herre 1949). Given the thresholds for the second and third molars determined in this study, these larger prehistoric specimens would still be classified as wild boar. Granted the existence of this complicating premise, however, transitional specimens originating from this larger wild ancestral phenotype could also possibly be identified as wild boar.

The morphological surrogate for the wild-domestic transitional form used in the present study (wild boar × feral swine hybrids) may not be a truly intermediate form. Although positioned correctly in the transitional size sequence (i.e., being smaller than the Eurasian wild boar, but larger than the prehistoric domestic morphological surrogate), the molar lengths and widths were closer to those of the domestic surrogate and were not truly intermediate in size. This is perhaps a result of the hybrid populations used in this study being predominantly feral swine in ancestry (Mayer and Brisbin 1991). The reduced contribution of the wild boar founding stock has produced a population which is morphologically more like the feral end of the hybrid spectrum in terms of molar morphology.

In some instances, a specific archaeological investigation may necessitate the identification of "culturally domesticated" (i.e., tamed/captive) wild ancestors versus truly domestic individuals. Unfortunately, the data suggest that nothing more than little to no distinguishable morphological changes would occur for many years following the first efforts at domestication of wild

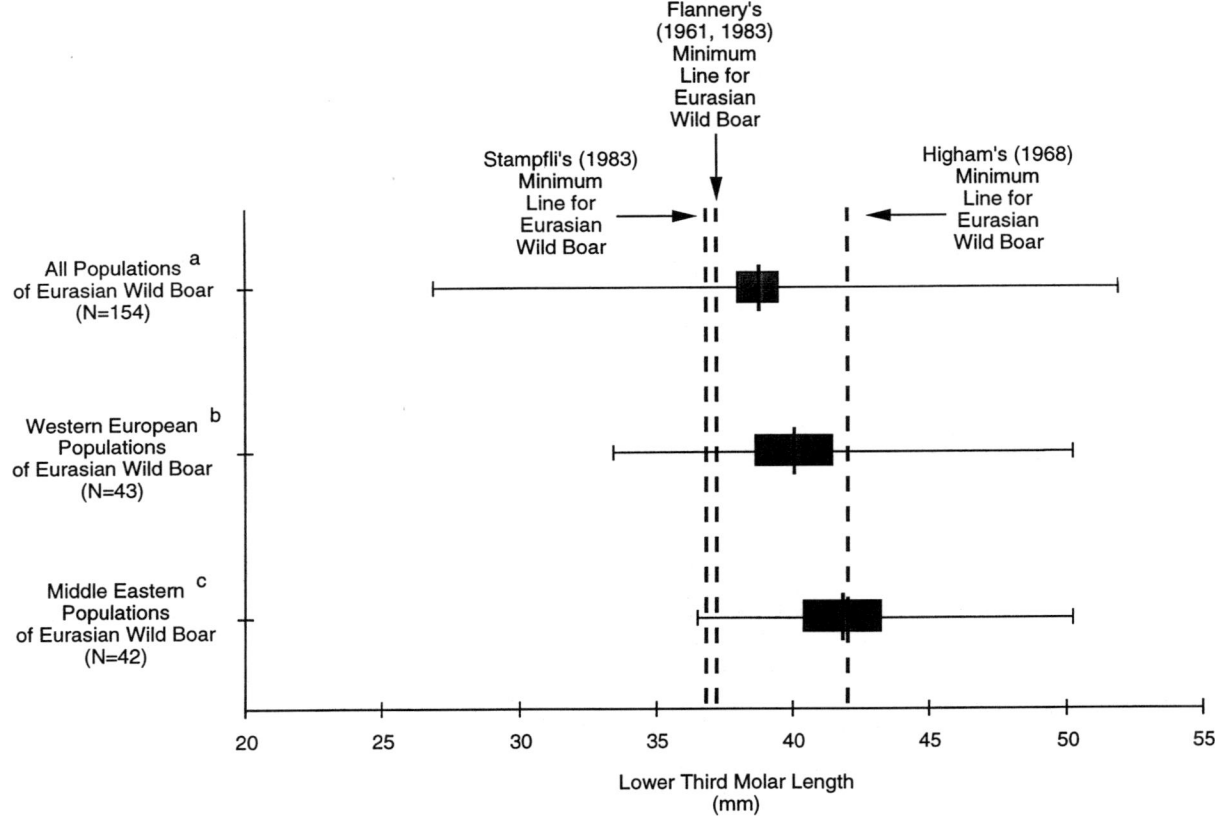

Fig. 5. Comparisons of lower third molar lengths between samples of all populations, western European populations, and Middle Eastern populations of Eurasian wild boar versus the minimum lines for Eurasian wild boar as given by Stampfli (1983), Higham (1968), and Flannery (1983). The bold vertical line, shaded box, and smaller vertical lines at the ends of the horizontal line represent the mean, 95% confidence interval, and observed range, respectively. The samples depicted were all measured in the present study and consisted of the following: (a) specimens of all Eurasian wild boar; (b) specimens of *Sus scrofa attila*, *S. s. majori*, and *S. s. scrofa*; and (c) specimens of *S. s. attila*, *S. s. davidi*, *S. s. lybicus*, and *S. s. nigripes*.

individuals. Within samples of recent captive wild boar, morphological analyses indicate that no significant quantifiable differences occur even after a number of generations. Based on specimens of zoo wild boar, the cranial morphology continues to be uniform, with such specimens still being classified in canonical variates analyses as Eurasian wild boar (Mayer and Brisbin 1991). Therefore, recently domesticated or tame wild boar living in the confined situation of an agricultural society may not be morphologically discernible from truly wild individuals being harvested solely under a strict hunting regime. Because of this fact and the lack of definitive archaeological evidence, the actual time period between initial domestication and a resultant response in the form of an observable morphological change in molar characteristics remains unknown.

Molar size alone cannot be used as the sole criterion for establishing the practice of an agrarian rearing (with selective breeding) of domestic swine for a given archaeological site or prehistoric society. The establishment of feral swine populations around some ancient settlements could have conceivably occurred given the early use of free-ranging husbandry practices for this species in some areas (Clutton-Brock 1981; Zeuner 1963). Jarman (1971) discussed the impossibility of detecting early feral individuals and the misleading interpretations that may result when using wild vs. domestic characters to distinguish between hunter-gatherer vs. agrarian-production societies. For example, even if such prehistoric societies had been provided with derived domestic swine through trade, it is possible that these animals could have been released to forage on their own until individual animals were harvested as needed by the human owners. The actual rearing practices used in such a scenario could have been minimal, and not comparable to other more strict agrarian-based social systems during the same his-

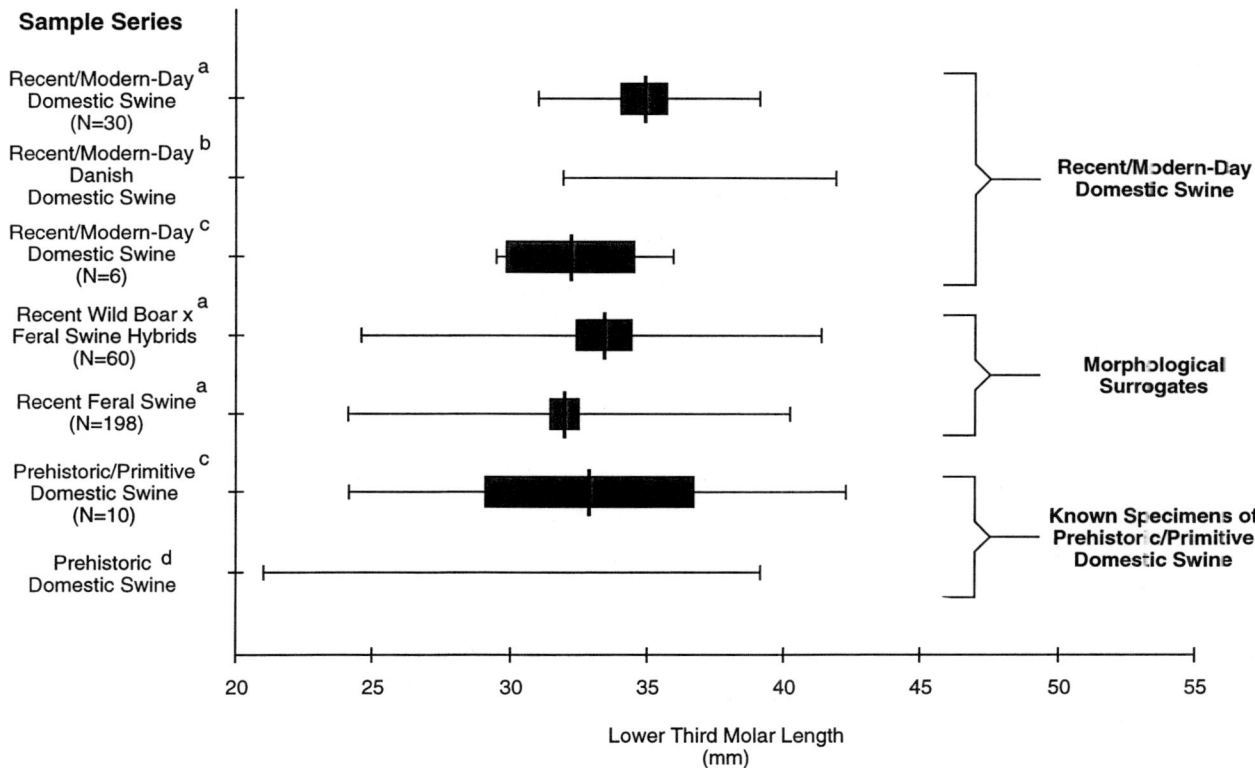

Fig. 6. Comparisons of lower third molar lengths among various samples of recent/modern-day domestic swine, the two morphological surrogates used in this study, and known specimens of prehistoric/primitive domestic swine. The sample conventions follow those defined for Figure 5. The sources of the sample series were as follows: (a) the present study; (b) taken from Higham (1968), sample size, mean, and standard deviation unknown; (c) taken from Pira (1909); and (d) taken from Nanninga (cited in Stampfli 1983), sample size, mean, and standard deviation unknown.

torical period. Further, the potential for free-ranging domesticates to hybridize with local wild boar could serve to further mask the presence of derived domestic swine and decrease the value of molar size as a parameter for identifying domesticates (Bogucki 1989; see also discussion in Redding and Rosenberg, this volume).

The most accurate (i.e., with the lowest percentage of incorrect classifications) measurement determined in the present study was upper second molar width, followed by the second cusp row width of the upper third molar. The best of the remaining measurements had incorrect percentages of only slightly less than twice that of the upper second molar length. Both Payne and Bull (1988) and Kuşatman (1992) pointed out the value of molar widths over lengths in looking for separations between wild and domestic populations of swine. This is attributed to the low overall variation, low sexual dimorphism, and low age-related variation exhibited by the molar widths (Kuşatman 1992; Payne and Bull 1988).

Comparisons with existing wild-domestic thresholds of third molar crown length reveal that some are reasonable (e.g., Flannery 1961, 1983; Stampfli 1983), while others (e.g., Higham 1968) are overly conservative (Fig. 5). Higham's (1968) minimum line for wild boar was greater than the mean of all of the subspecies found in and around the geographic areas where his study sites in Europe were located (i.e., *S. s. attila*, *S. s. majori* and *S. s. scrofa*). In fact, his minimum line is above even the upper limit of the collective 95% confidence interval for these subspecies.

Higham's and Flannery's methods provide a single threshold value above which an individual would be classified as a wild boar and below which a domestic *S. scrofa*. Stampfli's (1983) method uses a revised wild boar minimum observation (largely based on the same specimens used by Flannery [1961, 1983]), and combines that with a maximum observation taken from Nanninga (cited in Stampfli 1983). Stampfli's (1983) wild boar minimum value is less than the maximum domestic value, thereby creating a zone of overlap within which a specimen could be either type of *S. scrofa*. Our study has two threshold values similar to Stampfli's;

however, the wild and domestic threshold values do not overlap for any of the molar measurements studied. The specimens which fall between the two values would remain as unknowns, due to the inability to accurately identify these as either wild or domestic in origin.

Similar to the subspecific refinements made to the wild-domestic thresholds as depicted in Figure 5, it would also be important to set specific thresholds which consider only the molar sizes of the particular subspecies endemic to the area of the archaeological site in question, as well as those found in adjacent regions. Broader size variation would result if it is considered that subspecies found in such adjacent sites could have been translocated to the region of the archaeological site. For example, the area of the Fertile Crescent is located within the range of the subspecies *S. s. attila*. However, if one were also to consider the adjacent subspecies (*S. s. davidi*, *S. s. lybicus*, and *S. s. nigripes*), then a greater size variation would be expected as a result of possible translocations. In fact, both Flannery (1961, 1983) and Stampfli (1983) used specimens of *S. s. attila* and *S. s. lybicus* in their respective studies.

Comparisons between the two morphological surrogates and other samples of prehistoric and modern-day domestic swine provided mixed results. Stampfli's use of Nanninga's maximum for the lower third molar length of prehistoric domestic swine (cited in Stampfli 1983) compares favorably to the various samples of primitive and modern domestic swine and the two morphological surrogates used in the present study (Fig. 6). Again, a few extreme observations were above the threshold line; however, most would be correctly identified as domestics. At the same time, a few of Nanninga's (Stampfli 1983) specimens were smaller than any of the other data sets illustrated in the present study or in other studies of prehistoric domestic swine (e.g., Higham 1968). It should also be noted that none of the calculated upper limits of the 95% confidence intervals were equal to or exceeded the lower limit of the wild boar 95% confidence interval as determined in our study.

The potential of accidental incorporation of modern-day domestic swine into archaeological sites raises the question of how this recent material compares to Eurasian wild boar and the two surrogates investigated in the present study. In comparing modern-day domestic swine molars with these three types, the upper limits of the 95% confidence intervals of the recent domestics were above those of surrogates and below those of wild boar for all of the molar measurements. These upper limits for each measurement (in mm) within the modern-day domestic sample were as follows:

upper second molar: 21.8
upper second molar width: 17.2
upper third molar length: 33.6
width of first cusp row of upper third molar: 19.9
width of second cusp row of upper third molar: 17.5
lower second molar length: 21.5
lower second molar width: 15.1
lower third molar length: 35.7
width of first cusp row of lower third molar: 17.3
width of second cusp row of lower third molar: 16.9.

Thus, any modern-day domestic swine material becoming accidentally incorporated into an archaeological deposit is not likely to be identified as a wild ancestor, but at worst as an unknown. Based on the modern-day domestic specimens used in the present study, less than 4% of the recent domestic material would be above the minimum wild threshold.

In choosing the use of minimum/maximum observations vs. the upper/lower limits of 95% confidence intervals, one must decide upon the type of information desired as an outcome of the analysis to be applied to archaeological material under study. In comparing the results of the methods of Flannery (1983) and Stampfli (1983) with the present study, the percentage of known wild boar specimens that were incorrectly identified as domestic swine was more than two to almost three times higher using Stampfli's and Flannery's methods, respectively, than using the wild-domestic thresholds developed in the present study. However, the overall percentage of specimens which could not be identified as one type or the other was higher in the present study. Therefore, if one needed to be able to identify every specimen as either wild boar or domestic swine, the present method would not accomplish that goal. On the other hand, if one needed to more accurately determine whether or not the population sample from a given archaeological site represented wild boar vs. domestic swine, then the thresholds described in the present study would be a more reliable and statistically sound method.

The differences in width allometry of the lower second molar are an aspect of the dental variation among the types of *S. scrofa* that has not been previously noted by archaeozoologists. The primary practical difficulty of this method is that it requires a series of these molars from an archaeological site. Minimum sample sizes would be determined by the ability to produce a significant regression of molar width and length. Individual or small numbers of these teeth would not be a sufficient sample size to enable a wild-domestic identification. However, for locations producing large series of pig molars, this method could be used in conjunction with the ten 95% confidence interval thresholds for the second and third molars to determine what type of *S. scrofa* was present at those locations.

Conclusions

Comparisons of the molar size differences shown by recent specimens of *S. scrofa* have generally validated the use of molar size criteria to distinguish Eurasian wild boars from smaller primitive domestics. The effects of sex and age on the molars were not found to be substantial enough to obscure the differences between wild vs. domestic forms. This further substantiates the robust nature of the wild-domestic thresholds determined for use in this identification technique. Of the ten measurements included in the present study, upper second molar width was found to provide the lowest percentage of incorrect classifications and therefore the best potential for identifying wild vs. domestic forms accurately. On the same basis, the least useful measurement was lower second molar length. For the upper third molar, second cusp row width was a better discriminator than crown length, with the reverse relationship being found for the lower third molar. In both second molars, cusp row width was a more accurate discriminator than crown length. Analysis of Covariance indicated that the allometric relationship of lower second molar width differed between wild boar and the morphological surrogates. Given a sufficient sample size of *S. scrofa* lower second molars from an archaeological site, this would provide additional information that could increase the usefulness of this tooth for differentiation of individuals into wild boar or domestic swine morphotypes.

Based on 95% confidence intervals, the application of the thresholds developed during the present study would result in fewer incorrect identifications of pure Eurasian wild boar as early domestic swine. The primary shortfall of these thresholds would be the potential for a higher percentage of specimens to be identified as unknowns.

In spite of the conclusion from the present study validating the use of molar length and width to distinguish wild vs. domestic swine, the application of methods based on this size relationship should be undertaken with care. Tooth size alone is minimal evidence at best in trying to distinguish between hunting versus agrarian prehistoric societies or in determining wild-living vs. captive-reared specimens of swine. Molar size is an important small part of the domestication puzzle, but it is far from being either the complete picture or an infallible basis for identifying wild ancestors from truly domestic forms of swine.

Acknowledgments

We thank the following people for their efforts and assistance in the field collection of specimens: D.W. Baber, F. Bridges, K.O. Butts, M.B. Caudell, A.W. Conger, R.T. Hoppe, P.E. Johns, R.A. Kennamer, S. Kile, E. Rollor, R.F. Smart, D.E. Stallknecht, the late J.W. Reiner, Sr., M.J. Vargo, and E.T. West. Field collections of material were also undertaken in collaboration with personnel of the Southeastern Cooperative Wildlife Disease Study. We are indebted to the various institutions, curators, and staff who generously made the *S. scrofa* specimens in their care available for our examination. We would especially like to thank the following curators for the additional assistance which they provided: S. Anderson (American Museum of Natural History), P.W. Collins (Santa Barbara Museum of Natural History), P.W. Freeman (Field Museum of Natural History), H.H. Genoways (Carnegie Museum of Natural History), D.G. Huckaby (California State University), J.A.W. Kirsch (Museum of Comparative Zoology), J.L. Patton (Museum of Vertebrate Zoology), and A.I. Roest (California Polytechnic State University). F.D. Martin, U. Arbarello, and an anonymous reviewer critically read earlier drafts of the manuscript. This study was supported by funding provided by the U.S. Department of Energy to the University of Georgia under financial assistance award no. DE-SC09-96SR18546, and the Westinghouse Savannah River Company under contract DE-AC09-96SR18500.

References

Ammon, R. 1938. Abstammung, Arten, und Rassen der Wildschweine Eurasiens. *Zeitschrift für Tierzuchtung und Zuchtbiologie* 40:49–88.

Amschler, J.W. 1939. Tierreste der Ausgrabungen von dem "Grossen Konigshugel" Shah Tepe, in Nord-Iran. In *Reports from the Scientific Expedition to the North-Western Provinces of China under the Leadership of Dr. Sven Hedin, the Sino-Swedish Expedition*, Publ. 9, Vol. 7, Pt. 4., pp. 35–129. Bokforlags Aktieblaget Thule, Stockholm.

Briedermann, L. 1970. Zum Korper- und Organwachstum des Wildschweines in der Deutschen Demokratischen Republik. *Archiv für Forstwesen* 19: 401–420.

Brisbin, I.L., Jr. 1989. Feral Animals and Zoological Parks: Conservation Concerns for a Neglected Component of the World's Biodiversity. In *Regional Conference Proceedings of the American Association of Zoological Parks and Aquariums*, pp. 523–529. Wheeling, WV.

——— 1990. A Consideration of Feral Swine (*Sus scrofa*) as a Component of Conservation Concerns and Research Priorities for the Suidae. *Bongo* 18: 283–293.

Bogucki, P. 1989. The Exploitation of Domestic Animals in Neolithic Central Europe. In *Early Animal Domestication and Its Cultural Context*, ed. P.J.

Crabtree, D. Campana, and K. Ryan, pp. 119–134. MASCA Research Papers in Science and Archaeology, Supplement to Vol. 6. University of Pennsylvania Museum, Philadelphia.

Bökönyi, S. 1969. Archaeological Problems and Methods of Recognizing Animal Domestication. In *The Domestication and Exploitation of Plants and Animals*, ed. P.J. Ucko and G.W. Dimbleby, pp. 219–229. Aldine, Chicago.

—— 1974. *History of Domestic Mammals in Central and Eastern Europe*. Akademini Kiado, Budapest.

Bolomey, A. 1973. The Present Stage of Knowledge of Mammal Exploitation During the Epipalaeolithic and the Earliest Neolithic on the Territory of Romania. In *Domestikationsforschung und Geschichte der Haustiere*, ed. J. Matolcsi, pp. 197–203. Akadmiai Kiado, Budapest.

Case, T.J. 1978. A General Explanation for Insular Body Size Trends in Terrestrial Vertebrates. *Ecology* 59: 1–18.

Chaplin, R.E. 1969. The Use of Non-morphological Criteria in the Study of Domestication from Bones Found in Archaeological Sites. In *The Domestication and Exploitation of Plants and Animals*, ed. P.J. Ucko and G.W. Dimbleby, pp. 231–245. Aldine, Chicago.

Clutton-Brock, J. 1981. *Domesticated Animals from Early Times*. University of Texas Press, Austin.

Driesch, A. von den. 1976. *A Guide to the Measurements of Animal Bones from Archaeological Sites*. Peabody Museum Bulletin No. 1, Harvard University. Cambridge.

Epstein, H. 1971. *The Origin of the Domestic Animals of Africa*. 2 vols. Rev. ed. Africana Publishing Corp., New York.

Flannery, K.V. 1961. Skeletal and Radiocarbon Evidence of the Origins of Pig Domestication. M.A. thesis, University of Chicago.

—— 1983. Early Pig Domestication in the Fertile Crescent: A Retrospective Look. In *The Hilly Flanks. Essays on the Prehistory of Southwestern Asia Presented to Robert J. Braidwood, November 15, 1982*, ed. T.C. Young, Jr., P.E.L. Smith, and P. Mortensen, pp. 163–188. Studies in Ancient Oriental Civilization 36. Oriental Institute, University of Chicago. Chicago.

Foster, G.B. 1964. Evolution of Mammals on Islands. *Nature* 202:234–235.

Grigson, C. 1969. The Uses and Limitations of Differences in the Absolute Size in the Distinction Between the Bones of Aurochs (*Bos primigenius*) and Domestic Cattle (*Bos taurus*). In *The Domestication and Exploitation of Plants and Animals*, ed. P.J. Ucko and G.W. Dimbleby, pp. 277–294. Aldine, Chicago.

—— 1982a. Sex and Age Determination of Some Bones and Teeth of Domestic Cattle: A Review of the Literature. In *Ageing and Sexing Animal Bones from Archaeological Sites*, ed. B. Wilson, C. Grigson, and S. Payne, pp. 7–23. BAR, Oxford.

—— 1982b. Sexing Neolithic Domestic Cattle Skulls and Horn Cores. In *Ageing and Sexing Animal Bones from Archaeological Sites*, ed. B. Wilson, C. Grigson, and S. Payne, pp. 25–35. BAR, Oxford.

Groves, C.P. 1981. *Ancestors for the Pigs: Taxonomy and Phylogeny of the Genus Sus*. Tech. Bull. 3, Dept. of Prehistory, Res. School of Pacific Studies, Australian National University, Canberra.

Harrison, D.L. 1968. *The Mammals of Arabia*. Vol. 2, *Carnivora, Artiodactyla, Hyracoidea*. Ernest Benn, London.

Hell, P., and L. Paule. 1983. Systematische Stellung des Westkarpatischen Wildschweines *Sus scrofa*. *Acta Scientiarum Naturalium (Brno)* 17(3): 1–54.

Heptner, V.G., A.A. Nasimovic, and A.G. Bannikov. 1966. *Die Saugetiere der Sowjetunion*, Band I. Paarhufer und Unpaarhufer. VEB Gustav Fisher Verlag, Jena.

Herre, W. 1949. Betrachtungen uber Vorgeschichte Wildschweine Mitteleuropas. *Verhandlungen der Deutschen Zoologischen Gesellschaft in Erlangen, 1948*:324–333.

Herre, W., and M. Rohrs. 1990. *Haustiere—Zoologische Gesehen*. G. Fischer, Stuttgart.

Higham, C. 1968. Size Trends in Prehistoric European Domestic Fauna, and the Problem of Local Domestication. *Acta Zoologica Fennica* 120:1–21.

Jarman, M. 1971. Culture and Economy in the North Italian Neolithic. *World Archaeology* 2:255–265.

Keller, C. 1902. *Die Abstammung der Altesten Haustiere*. Verlag von Fritz Amberger Vorm. David Burkli, Zurich.

Kelm, H. 1938. Die Postembryonale Schadelentwicklung des Wild- und Berkshire-Schweines. *Zeitschrift fur Anatomie und Entwicklungsgeschichte* 108:499–559.

Koslo, P.G. 1975. *Dikij Kaban*. Izd. Urodzaj., Minsk.

Kowalski, K. 1976. *Mammals: An Outline in Theriology*. Panstwowe Wydawnictwo Naukowe, Warsaw.

Kurten, B. 1968. *Pleistocene Mammals of Europe*. Weidenfeld and Nicolson, London.

Kuşatman, B. 1992. *The Origins of Pig Domestication with Particular Reference to the Near East*. Ph.D. dissertation, University College, London.

Lawrence, B. 1980. Evidences of Animal Domestication at Çayönü. In *The Joint Istanbul-Chicago Universities Prehistoric Researches in Southeastern Anatolia*, ed. H. Çambel and R. Braidwood, pp. 285–308. Istanbul University Faculty of Letters Publication No. 2589, Istanbul.

Lowe, V.P.W., and A.S. Gardiner. 1976. A Re-Examination of the Subspecies of Red Deer (*Cervus elaphus*) with Particular Reference to the Stocks in Britain. *Journal of Zoology* 174:185–201.

Mayer, J.J., and I.L. Brisbin, Jr. 1988. Sex Identification of *Sus scrofa* Based on Canine Morphology. *Journal of Mammalogy* 69:408–417.

——— 1991. *Wild Pigs in the United States: Their History, Morphology, and Current Status.* The University of Georgia Press, Athens.

——— 1993. Distinguishing Feral Hogs from Introduced Wild Boar and Their Hybrids: A Review of Past and Present Efforts. In *Feral Swine: A Compendium for Resource Managers*, ed. C.W. Hanselka and J.F. Cadenhead, pp. 28–49. Texas Agricultural Extension Service, Kerrville, TX.

——— 1995. Feral Swine and Their Role in the Conservation of Global Livestock Genetic Diversity. In *Proceedings of the Third Global Conference of Domestic Animal Genetic Resources*, ed. R.D. Crawford, E.E. Lister, and J.T. Buckley, pp. 175–179. Rare Breeds International, Kingston, Ontario.

Oliver, W.L.R., and I.L. Brisbin, Jr. 1993. Introduced and Feral Pigs: Problems, Policy, and Priorities. In *Pigs, Peccaries and Hippos: Status Survey and Conservation Action Plan of the IUCN World Conservation Union*, ed. W.L.R. Oliver, pp. 179–191. International Union for the Conservation of Nature and Natural Resources, Gland.

Oliver, W.L.R., I.L. Brisbin, Jr., and S. Takahashi. 1993. The Eurasian Wild Pig (*Sus scrofa*). In *Pigs, Peccaries and Hippos: Status Survey and Conservation Action Plan of the IUCN World Conservation Union*, ed. W.L.R. Oliver, pp. 112–121. International Union for the Conservation of Nature and Natural Resources, Gland.

Payne, S., and G. Bull. 1988. Components of Variation in Measurements of Pig Bones and Teeth, and the Use of Measurements to Distinguish Wild from Domestic Pig Remains. *Archaeozoologia* 2(1, 2): 27–66.

Pira, A. 1909. Studien zur Geschichte der Schweinrassen Insbesondere Derjenigen Schwedens. *Zoologische Jahrbuch, Supplement* 10(2): 233–426.

Reed, C.A. 1969. Animal Domestication in the Near East. In *The Domestication and Exploitation of Plants and Animals*, ed. P.J. Ucko and G.W. Dimbleby, pp. 361–380. Aldine, Chicago.

Romic, S. 1975. Body Dimensions of Wild Boar. *Conspectus Agriculturae Scientificus* 34(4): 13–24.

Rutimeyer, L. 1862. Die Fauna der Pfahlbauten der Schweiz. *Neue Denkschriften der Schweizer Gesellschaft der Naturwissenschaft Zurich* 19:1–248.

SAS Institute Inc. 1989. *SAS/STAT User's Guide*. Version 6, 4th ed., Vol. 2. SAS Institute Inc., Cary, NC.

Spitz, F., G. Valet, and I.L. Brisbin, Jr. 1998. Variation in Body Mass of Wild Boars from Southern France. *Journal of Mammalogy* 79(1): 251–259.

Stampfli, H.R. 1983. The Fauna of Jarmo with Notes on Animal Bones from Matarrah, the Amouq, and Karim Shahir. In *Prehistoric Archaeology Along the Zagros Flanks*, ed. L.S. Braidwood, R.J. Braidwood, B. Howe, C.A. Reed, and P.J. Watson, pp. 431–483. Oriental Institute Publication Vol. 105, University of Chicago, Chicago.

Stein, G. 1989. Strategies of Risk Reduction in Herding and Hunting Systems of Neolithic Southeast Anatolia. In *Early Animal Domestication and Its Cultural Context*, ed. P.J. Crabtree, D. Campana, and K. Ryan, pp. 87–97. MASCA Research Papers in Science and Archaeology, Supplement to Vol. 6. University of Pennsylvania Museum, Philadelphia.

Teichert, M. 1969. Osteometrische Untersuchung zur Berechnung der Widerristhohe bei vor- und Fruhgeschichtliche Schweinen. *Kuhn-Archiv* 83:237–292.

Van Vuren, D., and P. Hedrick. 1989. Genetic Conservation in Feral Populations of Livestock. *Conservation Biology* 3:312–317.

Zeuner, F.E. 1963. *A History of Domesticated Animals*. Harper and Row, New York.

EARLY PIG HUSBANDRY IN SOUTHWESTERN ASIA AND ITS IMPLICATIONS FOR MODELING THE ORIGINS OF FOOD PRODUCTION

Michael Rosenberg

University Parallel Program, University of Delaware, 333 Shipley St., Wilmington, DE 19801

Richard W. Redding

Museum of Anthropology, 1109 Geddes Avenue, University of Michigan, Ann Arbor, MI 48109-1079

Introduction

Approximately thirty years ago the first generation of non-progressive, *causal* explanations for the origins of food production first appeared, spearheaded by what soon came to be called the Binford-Flannery model (see Wright 1971; Binford 1968; Flannery 1969, 1973). It dealt specifically with agricultural origins in southwestern Asia, but it embodied general mechanisms revolving around population pressure that were quickly extrapolated to other parts of the world as well (e.g., Meyers 1971). These first generation causal models were justifiably criticized on a number of grounds (e.g., Cowgill 1975; Bronson 1975; Harris 1978; Ellen 1982; Rindos 1984; Price and Brown 1985; Rosenberg 1990; Henry 1991; Hayden 1992) and have generally been replaced by a second generation of models that more often than not seek to explain the origins of food production by reference to processes other than population pressure.

These second generation models are generally more detailed in their proposed workings than were the first generation models they supplanted. However, most are still based on simple unilinear causal chains. Recent discoveries at Hallan Çemi[1] in eastern Turkey, including evidence for the apparent early domestication of pigs at that site (see Redding and Rosenberg, this volume), indicate that these second generation models, in their unilinear reasoning, are in their own ways as limited as the first generation models they replaced.

This paper summarizes the Hallan Çemi data and examines the somewhat simplistic assumptions integral to the large majority of current models seeking to explain the origins of food production in either regional or general terms. It argues that the origins of food production is a more complex process than allowed for by most current models. It further argues that there are multiple levels of causation integral to the process, with both socio-demographic and socio-political forces operating in different capacities at two of these levels.

Hallan Çemi

Hallan Çemi is a relatively small (under 0.5 ha) aceramic site in the southern-facing foothills of the Taurus mountains in eastern Turkey, which was apparently occupied for several hundred years toward the end of the eleventh millennium B.P., uncalibrated (Rosenberg 1994a; Rosenberg et al. 1995). It is situated at an elevation of 640 m on the west bank of the Sason Çayı (a tributary of the Batman River and Tigris, respectively), which cuts through these foothills to form a small valley. Extensive oak forests currently cover both the higher elevations and the unfarmed portions of the lower elevations in this foothill zone.

Several kilometers to the north of this valley, the snow-capped peaks of the Sason Dağları rise well above the tree line. To the south, the foothills terminate at an expanse of relatively flatter terrain that slopes gently to the south for ca. 30 km, until cut by the Tigris. This flatter terrain is cut by the wide floodplain of the Batman River[2] and is currently devoid of forests. Annual precipitation in this entire region is currently sufficient to support dry cereal farming.

Based on a preliminary analysis of 125 wood charcoal fragments[3] (derived from 3 of the float samples), *Fraxinus*, *Quercus*, *Prunus*, *Pistacia*, and *Salix* or *Populus* are present in the site deposits. Buckthorn (cf. *Frangula alnus*) may also be present. The *Salix/Populus* charcoal suggests the proximity of riverine forests to the site during the period of its occupation, while the other species are consistent with a mixed oak forest essentially similar to that which presently covers the unfarmed areas in the vicinity of the site.

The architectural features at the site encompass both stone structures, including two particularly large semi-subterranean examples in the uppermost building level (see Rosenberg 1994a: Figs. 7, 8), and the indistinct remains of other structures made of more perishable materials. However, Hallan Çemi was certainly not a particularly large community. The number of clearly identifiable (i.e., stone) structures is relatively small, at no more than five per several hundred square excavated meters in each level. Even if we accept all the possible other structures suggested by plaster patches, post holes, and traces of what may be collapsed roofing material, it is still hard to justify the conclusion that the site contained more than 15 to, at the very most, 20 structures at any one time. Thus, it seems that Hallan Çemi was inhabited by a community only slightly larger than that thought to characterize mobile hunting-gathering bands (cf. Lee and DeVore 1968).

Hallan Çemi was clearly occupied year-round. In addition to substantial architectural remains, the plant and animal remains provide various lines of evidence indicating that Hallan Çemi was home to a sedentary society. Of these, the most direct evidence is in the growth bands on the clam shells (*Unio tigridus*) found at the site. Of the 63 clam samples with ventral margins sufficiently intact to permit analysis,[4] 16% were harvested during the period of slow growth (i.e., summer), 43% were harvested early into the rapid-growth phase, 19% were harvested well into the rapid-growth phase, and 16% were harvested near the end of the rapid-growth phase.[5] Thus, clams were apparently gathered by Hallan Çemi's inhabitants over the course of the entire year.

There is also evidence for a degree of cultural complexity at Hallan Çemi that is not usually characteristic of mobile hunter-gatherers. For example, the entrances to obvious structures in all levels of the site generally face away from the central (i.e., communal) activity area (Rosenberg et al. 1995: Fig. 1). As noted by Yellen (1985, 1990), with respect to similar layouts in the camps of some !Kung groups that have recently begun to practice pastoralism, this is a departure from the layout characteristic of the camps of mobile hunter-gatherers. Yellen suggests that this change relates to the abandonment of generalized reciprocal sharing at the community level and its replacement by the more restrictive exchange networks that develop in response to the institutionalization of ownership (see also Byrd 1994:649ff.).

Public buildings are also present at the site in at least the uppermost building level. These are distinguished by their size, floor treatment, the general absence of artifact types usually associated with food processing (e.g., grinding slabs), and distinctive architectural features such as stone platforms/benches placed along the walls. One of the two such structures in the uppermost building level also contained an aurochs skull, which (judging from its condition, position, etc.) once hung on the wall facing the entrance (see Rosenberg 1994a). It constitutes the *only* aurochs remains thus far found in the entire faunal assemblage. Moreover, the distribution of several imported materials (i.e., copper ore and obsidian) at the site suggests that, whatever else their social role may have been, the groups associated with these public structures were involved in the trade of these commodities (Rosenberg, Redding, and Akin n.d.).

There is also evidence at this site for the conspicuous consumption of foodstuffs—in other words, feasting (see also Hayden 1995). This evidence takes the form of a large central activity area with inordinately dense deposits of fire-cracked rocks and animal bones that are often still articulated at the level of a complete animal part. A linear arrangement of three sheep crania (with horns still attached) was also found in this area (see Rosenberg and Davis 1992; Rosenberg 1994a). The evidence for feasting also takes the form of numerous decorated stone bowls and sculpted stone pestles. These are typically made of the same fine-quality stone and to the same general scale. Together they suggest the formalized preparation and consumption of foodstuffs (Rosenberg and Redding 1995; Rosenberg, Redding, and Akin n.d.).

Lastly, there is a series of small notched stones. With one exception, this type is consistently made of a relatively soft, micaceous, metamorphic rock (perhaps a schist). Though the full form is problematic, all the pieces (with the same exception) exhibit certain features. Those are: a lenticular cross section with a maximum width of no more than 1.5 cm and maximum height of ca. 3 cm; a length of at least 15 cm, tapering down to either a convex or dimpled tip; and, a variably spaced series of sharp notches cut into one or both edges, usually quite neatly (see Rosenberg 1994a: Fig. 12). These notches range from one to eight in number on the examples at hand. To the naked eye there is no evidence of wear within the notches or elsewhere on these objects; the notches were simply cut into the stone. Given the lack of any use wear on these objects (despite the softness of the stone) and the relatively high degree of standardization within the type in all respects save the number of notches, these notched batons can most readily be construed to represent formal tallies of some sort (see Rosenberg et al. 1995). If so, whatever was being tallied was arguably socially, economically, or politically important enough to record.

The point here revolves not around what specifically was being recorded, as that is probably not knowable archaeologically. Rather, it revolves first around the fact that the social recognition of *any* socially, economically, or politically significant act is studiously avoided by mobile hunter-gatherers (cf. Lee 1969). Thus, these notched batons again suggest the existence of a degree of social complexity not typical of mobile hunter-gatherers. Secondly, the social recognition of any such acts implies achieved status, and hence the absence of social stratification.

The economy of the Hallan Çemi's inhabitants was essentially based on hunting and gathering. Almonds, pistachios, and pulses were apparently the most intensively utilized plant resources. Cereal grasses do not appear to have been utilized at all (see Rosenberg et al. 1995). This is particularly noteworthy, as cereal grass exploitation is central to most current models attempting to explain the beginnings of both sedentism and agriculture in southwestern Asia at the end of the Pleistocene (e.g., Henry 1989, 1991; McCorriston and Hole 1991).

Wild sheep and red deer were the two most intensively exploited animal species at Hallan Çemi, respectively constituting ca. 40%[6] and 25% of the mammalian remains in the faunal assemblage. Pigs were a significantly less important component of the diet in purely quantitative terms, constituting only ca. 17% of the mammalian remains. However, judging from the molar sizes (cf. Flannery 1982), butchering patterns, sex ratios, and survivorship curves (see Redding 1994; Rosenberg et al. 1995:5; see also Redding and Rosenberg, this volume), the site's inhabitants were engaging in some degree of pig husbandry by at least the last building level—the end of the eleventh millennium B.P. (uncalibrated). This is significant, as it was previously thought that the economically more important ovicaprids were the earliest animal domesticates in southwestern Asia (e.g., Hole 1989).

Lastly, there is some limited evidence for both territoriality and the mode of territorial organization among the inhabitants of the upper Tigris drainage during the round house horizon.[7] The Batman drainage has been systematically and intensively surveyed over the course of several seasons (Algaze et al. 1991; Rosenberg and Togul 1991). Only two stratified early aceramic sites are known to exist within that drainage, Hallan Çemi and Demirköy.[8]

Demirköy is situated on the west bank of the Batman River ca. 40 km down-river from Hallan Çemi. Survey collections made in 1989 and 1993 and soundings at this site conducted in 1997[9] yielded a lithic assemblage containing both scalene triangles of the type so common at Hallan Çemi and projectile points of early tenth millennium B.P. type (e.g., nemrik type points). Obsidian was significantly less common at Demirköy than at Hallan Çemi.

The lower frequency of obsidian and scalenes in the Demirköy assemblage and the presence in that assemblage of nemrik points combine to suggest that the occupation at Demirköy postdates the one at Hallan Çemi and likely represents a relocation down-river of the same community that had previously occupied Hallan Çemi.[10] The point is that during the entire round house horizon there was never more (or less) than a single resident community exploiting the Batman drainage, implying that this community's territory constituted the entire Batman drainage. Moreover, the accumulating results of surveys (e.g., Algaze et al. 1991; Benedict 1980:167: S63/26) and chance discoveries[11] suggest that virtually all significant tributary drainages along the upper Tigris were home to such early aceramic communities. Whether each of these was also home to only a single community remains to be demonstrated. However, it is worth noting that the equation of territory and natural drainage boundaries conforms to what is thought to be the norm for hunter-gatherers (e.g., see Jochim 1976).

Explaining the Origins of Food Production

The first generation of causal models[12] that attempted to explain the origins of food production in processual terms (e.g., Wright 1971; Binford 1968; Flannery 1969, 1973) were based largely on the proposition that it resulted from growth-fed population pressure. These models were ultimately subject to criticisms on two fronts.

Criticisms of the first type focused on the likelihood of growth-fed population pressure. Typically, critics focusing on this issue argued that there is ample evidence to indicate that hunter-gatherers regulate population levels to remain well below carrying capacity (e.g., Cowgill 1975; Bronson 1975; Harris 1978; Ellen 1982; Henry 1991). Alternately, they argued that there is little in the archaeological record to indicate significant population growth during the relevant critical intervals (e.g., Henry 1991; Hayden 1992).

Criticisms of the second type focused on the mechanics of these early population pressure models. Critics argued that population pressure *per se* cannot be said to directly cause anything but stress. Stress, they correctly observed, is not the same as innovation (e.g., see Price and Brown 1985; Rosenberg 1990, 1994b; see also Rindos 1984). Thus, these early population pressure models simply did not lay out the process whereby one ostensibly leads to the other.

Criticisms of the first type have been addressed elsewhere (Rosenberg n.d.; see also Keeley 1988) and will not be rebutted again in detail. It will suffice here to note two points. First, population growth is not the only conceivable form of population pressure (e.g., see Harris 1978). Second, there are sound evolutionary reasons to question the supposed ability of hunter-gatherers to regulate population in the absence of resource conditions (i.e., pressure) requiring them to do so (e.g., see Bates and Lees 1979; Rosenberg n.d.).

As for criticisms of the second type, the fact that the early population pressure models did not lay out the processual relationship between population pressure and innovation does not mean it cannot be done (e.g., see Rosenberg 1990, 1994b). The point is that these criticisms of the second type are valid only with respect to a particular early, dated, and rather simplistic view of population pressure (e.g., Binford 1968). Thus, they have little or no bearing on a more detailed and evolutionarily sound view of pressure-induced stress and its role in the evolution of culture (see Rosenberg 1990, 1994b).

In any case, in lieu of socio-demographic factors, most recent attempts to explain the origins of food production at either a regional or general level have focused on other mechanisms. In southwestern Asia, for example, it has variously been suggested that food production is: (1) the outcome of coevolutionary processes (Rindos 1984); (2) the outcome of innovative resource exploitation patterns that focused on cereals, encouraging sedentism and thus making food production inevitable (e.g., McCorriston and Hole 1991); (3) the outcome of innovative resource exploitation patterns that required sedentism, which then made food production inevitable (e.g., Henry 1989, 1991); and, (4) the outcome of socio-political competition that promoted intensified exploitation patterns (e.g., Hayden 1990, 1992). A similar range of explanations has been offered for the beginnings of food production in other areas as well (e.g., see Wills 1988; Higham and Maloney 1989; Cowan and Watson 1992; MacNeish 1992; Charles 1992).

However, with the exception of Hayden's feasting model (see also Charles 1992), to which we will return shortly, these various models all tend to embody a common assumption. This assumption is that food production develops in a more or less straightforward, linear fashion. That is, given a particularly important food resource, such as cereal grasses or ovicaprids, these models implicitly submit that food production is simply the culmination of *that* resource's evermore intensive exploitation. This progressively more intensive exploitation is made necessary or possible by whatever forces are said by the model to be at work. Thus, in addition to being unilinear, they are also unicausal, a point to which we will return later.

As the Hallan Çemi data make abundantly clear with respect to the beginnings of animal husbandry, an assumption of linearity is at best highly questionable. Simply because a particular food species became a domesticated staple of later economies that were based on food production does not mean that it necessarily figured prominently in the early stages of the process that produced those economic systems. This applies even if that resource was intensively exploited during these early stages. Thus, as regards the development of animal husbandry, despite the fact that pigs were a secondary animal resource at Hallan Çemi as compared to both ovicaprids and deer, it was pigs that were apparently first produced at that site. On further study, that also seems to have been the case at Çayönü (see Hongo and Meadow, this volume; see also Lawrence 1982).

This is presumably because pigs were both easier to domesticate and more productive to produce than ovicaprids at the earliest stages of the process (see Redding 1994; Rosenberg et al. 1995; see also Redding and Rosenberg, this volume). In eastern Anatolia, it was apparently only later that the lessons learned by working with pigs were applied to ovicaprids for reasons that are still not entirely clear (see Redding and Rosenberg, this volume).

Hayden's competitive feasting model (1992) is one of the few exceptions to this linear approach (see also Blanton and Taylor 1995). It proposes that the earliest produced foods are not dietary staples, but rather high-prestige foods. It further proposes that the initial impetus to produce food had little to do with diet and everything to do with the enhancement of prestige. Pigs certainly fall under the category of prestige food in many areas of the world (cf. Blanton and Taylor 1995), and the early husbandry of pigs at Hallan Çemi is thus consistent with Hayden's model; so is the apparent evidence for feasting at Hallan Çemi. However, there are theoretically troubling aspects to Hayden's model and it too is unicausal. Moreover, there are aspects of the Hallan Çemi faunal data which suggest that even Hayden's model does not do justice to the complexity of the processes leading to food production.

Pigs, Feasting, and Food Production at Hallan Çemi

Theoretical issues momentarily aside, the basic problem for Hayden's model is that, aside from the actual evidence for pig husbandry at Hallan Çemi, there is also evidence at Hallan Çemi for possible practices that are inconsistent with the implications of Hayden's model.

By definition, food production involving any type of plant or animal resource requires a practitioner to forego the immediate consumption of some part of that resource in order to enhance its future availability.[13] Food production is thus a type of delayed-return system. Based on the range of current models for the origins of food production, in a given society either social or caloric considerations can be said to take precedence in the decision-making process leading to surplus production. However, whichever it is, it must consistently do so. Thus, if social (as opposed to caloric) considerations were the primary factor in the decision to delay returns (i.e., produce surpluses), we can expect all evidence for delayed returns to somehow be relatable to social phenomena. In other words, if social considerations outweighed caloric considerations in the production of animal surpluses at Hallan Çemi, we can expect that all evidence for delayed (animal resource) returns will be relatable to activities (such as feasting) that function to enhance individual prestige. Conversely, any evidence for behavioral patterns producing delayed (animal resource) returns that do not directly or indirectly enhance the status of the individual practicing them can be construed to mean that the impetus to engage in them was more fundamentally dietary in nature. Moreover, that would imply that caloric considerations superseded social considerations at a general level. The data on the sheep remains from Hallan Çemi are, therefore, noteworthy.

As noted above, the sheep remains from Hallan Çemi are all of morphologically wild individuals and there is as yet no other evidence (size reduction, age/sex skewed survivorship curves, evidence of on-site butchering, etc.) to indicate that sheep husbandry was practiced by the site's inhabitants. Nevertheless, 68%[14] of the sheep remains from Hallan Çemi that could be sexed[15] were those of males, a highly skewed figure.[16] We suggest that this may indicate selective hunting (i.e., culling) of males as a simple means of resource conservation.[17]

Such a willingness by individuals to voluntarily delay resource returns (by occasionally refraining from killing female sheep to enhance future availability of sheep) obviously requires some cultural system to insure that the individuals making the behavioral investment will reap the long-term benefits of that behavior. First, at the group level, territorial ownership must exist; and second, such selective hunting must both benefit the group as a whole and not harm the individuals practicing it. The first is necessary to limit access to conserved resources to only those practicing conservation (i.e., territorial holders). The second is necessary to insure that some group members do not benefit excessively from the sacrifice of others.

With regard to the first prerequisite, one of us (Rosenberg 1990 and n.d.) has argued elsewhere that territoriality is a precondition for sedentism, such as exhibited by Hallan Çemi's inhabitants. Thus, it is not an issue in the matter at hand. Also, as noted earlier, there is some evidence for such territorial organization along the upper Tigris. However, the second prerequisite raises some interesting implications.

In *any* system of sharing, whether reciprocally generalized or redistributive (as in the context of feasting), individuals generally benefit calorically from any hypothetical conservation efforts to the precise degree that the group as a whole benefits. In this particular case, an individual who forgoes a specific kill for reasons of conservation will nevertheless tend to share in any enhanced future caloric return on that action. This is because the future kills will be shared by all in the same manner as the foregone kill would have been. However, if there are additional benefits (beyond caloric), such as enhanced prestige, which accrue only to the individual actually making the kill, a willingness to delay returns hinges on the likelihood of actually reaping the reward on one's investment.

The point is that since hunting always involves a high degree of chance, such assurance of delayed prestige return to the individual (as opposed to caloric return to the group) is virtually impossible and such conservation practices cannot be expected in situations where hunting success overtly contributes to the hunter's prestige.[18] An individual may occasionally be willing to forego a kill in order to enhance the future availability of meat, but only if making a kill has no added prestige value which is essentially lost forever once the decision is made to forgo that kill. Thus, the sheep data suggest that prestige considerations were either not factored into hunting decisions or were outweighed by caloric considerations. In either case, the suggestion is that caloric considerations took priority over social considerations in decisions concerning meat procurement at Hallan Çemi and that presumably applies to husbandry as well as hunting practices.

Moreover, given the reconstructed (i.e., small) size of the Hallan Çemi community and its implications for social structure (see Rosenberg, Redding, and Akin n.d.), it is likely that any feasting at that site was more for purposes of alliance building in the Yanomamo mold (cf. Chagnon 1983) than for purposes of prestige enhancement (in the Kwakiutl mold). Hayden (1996) appears to formally exclude such "reciprocal aid" feasts from his competitive feasting model. However, he has ventured that the desire by the group as a whole to impress potential allies would also be a powerful incentive to intensify production or (at a minimum) insure the availability of desirable foods (Hayden, pers. comm. 1997). Husbandry practices would certainly serve to accomplish either. The problem here is that, in the best documented case of such reciprocal aid feasting, animal husbandry does not appear to have been practiced to do so despite the frequent dearth of meat (see Chagnon 1983).

The data aside, there are also theoretical assumptions inherent in Hayden's model (as currently formulated) which ignore the fundamental fact that social payoffs, such as enhanced prestige, are dependent on *preexisting* social context. That is, there is a clearly documented tendency in societies that do not already produce surpluses to actively belittle individuals behaving in ways even remotely similar to those required by Hayden's model (cf. Lee 1969). Thus, any individual acting in the ways proposed by Hayden would find the social results to be counterproductive; they would actually lose prestige instead of gaining it, and so would not behave that way. Hayden's model presupposes the legitimacy of a behavior, whose legitimacy is dependent on the pre-existence of the very thing it is claimed to bring about. That is, the manipulation of individuals by others to produce surpluses can only occur in societies that already accept the production of surpluses as a legitimate endeavor. As discussed in detail elsewhere (Rosenberg 1994b:322ff.), that only comes about when forces come into play which raise the specter of penalties more intimidating than the loss of prestige for failure to share (i.e., the accumulation of surpluses). These forces are dietary and these penalties bear on matters of survival.

All this, however, is not to say that feasting could not have played a prominent role in the evolution of husbandry practices at Hallan Çemi and elsewhere. Any infrastructural innovation, such as animal husbandry, food storage, etc., that produces caloric surpluses for purposes of short-term subsistence risk management, will inevitably produce at least occasional surpluses that are truly surplus. To the degree such intermediate to long-term surpluses are a regular occurrence, they become a feature of the socio-economic landscape, the cultural playing field so to speak, and people will alter the structured agonistic "games" they play to take into account the new form of playing field. In other words, the existence of surpluses, produced originally for subsistence-related reasons, will soon prompt people to utilize those surpluses in ways beyond their original intended function. One such way could easily be as a medium for agonistic competition.

Such latent structural functions may quickly come to supersede the original subsistence function in the minds of the practitioners, to the point where the original manifest function for the underlying behavior (surplus production for subsistence reasons) is forgotten and surpluses are produced strictly for (social) *purposes* of competition.[19] In fact, the practical limits on pig herd size coupled with the virtual absence of such limits on potential ovicaprid herd size raises an interesting possibility. It is that the shift to ovicaprid husbandry in eastern Anatolia sometime following the abandonment of Hallan Çemi could very well have been prompted by nothing more than the desire to intensify animal production, perhaps for purely social purposes.

Proximate and Ultimate Causation

Linearity aside, the other problem with virtually all (including non-linear) current models for the origins of food production is that evolutionary causation is proposed to occur in only one dimension—selection, climatic change, dietary stress, social forces, etc.—when in dealing with the evolution of culture it almost certainly operates simultaneously in several dimensions. For example, at the historical (evolutionary) level we can legitimately talk about proximate causation, e.g., climatic change and its effects on the resource base, which make a previously impractical type of subsistence behavior possible. Explanations such as those offered by Henry (1989, 1991) and McCorriston and Hole (1991) would fall under this heading. At the same level, however, we can also legitimately talk about ultimate causation, e.g., selection for risk-minimizing behaviors in a given environmental context, such as sedentism, storage, food production. The explanations offered by Rindos (1984) and Redding (1988) would fall under this heading. However, when dealing with the evolution of culture, we can also talk about additional dimensions of proximate causation at the level of cultural process. Thus, we can discuss demographic stress and its effects on cultural innovation, the process that first produces risk-minimizing behaviors (among other things) to be selected for or against (see Rosenberg 1990, 1994b). We can also consider the latent (including social) functions a given behavioral pattern takes on over time, which may reinforce the willingness (or even become the primary motivation) to engage in the behavior.

After all, the evolutionary relationship between the production of cultural novelties (i.e., innovation) and selection is not, as is the case of biological novelties (i.e., mutations) and selection, a purely mechanistic process (cf. Rosenberg 1990, 1994b). If people do not *see* a particular innovative behavior as being somehow in their best interests, they will not engage in it. It will thus be unavailable for selection to operate on, *no matter how potentially advantageous it could have proven to be*. This is the process often referred to as cultural selection (e.g., see Boyd and Richerson 1985).[20]

Our point is not just that the evidence for early pig husbandry at Hallan Çemi indicates that existing models for the origins of animal husbandry in southwestern Asia are incomplete. Our further point is that the whole approach to the origins of food production *in general* has failed to appreciate the complexity of the process. This is indicated by not just the Hallan Çemi pig data, but by the botanical data as well. For example, it has typically been proposed that intensive cereal grass exploitation was central to the beginnings of sedentism in southwestern Asia. From that starting point, it is generally proposed (e.g., Rindos 1984; Henry 1989) that the production and later domestication of cereals was simply the end result of continued, evermore intensive exploitation.

At Hallan Çemi, cereals are conspicuously absent, indicating first of all that cereal grass exploitation *in and of itself* is not a necessary and sufficient condition for the beginnings of settled village life and food production. Moreover, recent studies (e.g., Miller 1996) raise questions as to whether even the cereal grasses commonly found at early Levantine sites, and cited as evidence for cereal grass exploitation at those sites, are in actuality food remains. Lastly, in the American Basin, the only well-documented ethnographic case of incipient cultivation involving grasses, the grasses in question were produced as supplements to the harvesting of (wild) nuts (see Steward 1938), in much the same fashion as pigs were apparently produced at Hallan Çemi to supplement the hunting of wild sheep and deer. The point is that there are valid grounds for believing that the beginning of agriculture is also a considerably more complex process than allowed for by current models. It is thus at least worth reconsidering the possibility that, perhaps even in the Levant, sedentism may have initially been based on the exploitation of nuts or some other resource, with the exploitation of cereals being supplementary to that resource and involving their production from the very beginning (e.g., see Rosenberg 1990, 1994b; Olszewski 1993).

Conclusion

In conclusion, implicit in many of the models seeking to explain the beginnings of food production and settled village life in the Near East is the assumption that the exploitation of the domesticated species known to be the staples of later Neolithic economies was central to the early development of those economies. As the Hallan Çemi data make quite clear, this was apparently not the case along the Taurus-Zagros flanks, as regards both the origins of agriculture and animal husbandry in that region. Moreover, the Hallan Çemi data, notably those pertaining to pigs and other elements of the faunal assemblage, indicate that the origins of animal husbandry was a considerably more complex process than allowed for by current models. They indicate that if we are ever to truly understand the process, causation must be addressed on multiple levels, not just a single one.

Notes

1. Funding and other support for the Hallan Çemi excavation was provided by the National Geographic Society, the National Science Foundation, Mobil Exploration Mediterranean Inc., the University of Delaware Research Foundation, the University of Delaware, the University of Pennsylvania, the American Research Institute in Turkey, and the University of Michigan.

2. See Algaze et al. 1991 for a more detailed description of this area's geomorphology.

3. We are indebted to Rowena Gale for providing these data.

4. We are indebted to Mr. Keith Doms (University of Delaware), who graciously provided us with the results. A radial edge was ground, polished and then etched with 5% HCl. This prepared edge was then examined with a 10x lens to determine the amount of growth that had taken place since the last slow (translucent) growth or summer band (see Clark 1979; Quitmeyer, Hale, and Jones 1985).

5. The estimate of where the "end" of the current year's rapid growth band would have been is based on the size of the previous year's rapid growth band.

6. Sheep/goat remains collectively constitute ca. 42% of the mammalian bone. Among those bones types where the difference between sheep and goats can be distinguished, sheep outnumber goats by about 18:1. On the assumption that the ratio of sheep to goats in these particular categories of bone are representative of the larger body of sheep/goat bone, sheep constitute ca. 40% of the mammalian bone in the faunal assemblage.

7. That is, the terminal Epipaleolithic and early aceramic Neolithic—the local equivalent of the late Natufian and PPNA in the Levant.

8. See Algaze et al. 1991:181–182 and Fig. 3 (where Demirköy is called Demirci Höyük). A third stratified site, related to the PPNB scatter at Nevala Denik (see Rosenberg and Togul 1991), is suspected to exist, but its existence remains to be confirmed. However, such a possible late aceramic site within the Batman drainage does not bear on what follows beyond perhaps extending the sequence of community relocations even further.

9. Funding for the 1997 excavation at Demirköy was provided by the National Science Foundation and the International Nut Council.

10. The presence of sickle hafts and blades at Demirköy (see Algaze et al. 1991: Figs. 3:6 and 4) and the absence of such arti-

facts at Hallan Çemi suggest that a shift to the exploitation of cereals may have played a role in that relocation.

11. These include Ambarköy Höyük (on the Ambarçay) and Hirbet Selim (in the Gercüş vicinity), discovered respectively by the staffs of the Diyarbakır and Mardin provincial archaeology museums.

12. As opposed to discovery models (e.g., Childe 1951).

13. For example, not killing and immediately eating an animal in order to breed it, so that both it and its offspring are available to eat in the future.

14. The exact ratio is 19 males to 9 females.

15. Sexing was based on the pubis, atlas, and axis only. Horn cores (which are not as reliable) were not used.

16. Assuming an even sex tertiary ratio in the population from which the sample was drawn, the probability of getting, by chance, 19 males in a sample of 28 individuals is .04358.

17. Such culling would have no impact on the reproductive rate.

18. This is simply a variant of the "prisoner's dilemma" puzzle of game theory in reiterated form (e.g., see Maynard Smith 1982:167ff., and Dawkins 1989:203ff. for discussions of "prisoner's dilemma"). Reciprocal failure to cooperate (in conservation) is ineffective as punishment for an initial failure to cooperate. Therefore, all else being equal, a rational individual has little choice but to defect (i.e., not conserve).

19. However, such surpluses, as long as they are produced, will continue to serve their original subsistence function, even if that function is later obscured by latent (social) functions. That is, they will still be available in time of need.

20. It is, incidentally, in this respect (as reinforcement) that we believe the triggering of a feasting model (based on the inception of surplus production for initial dietary reasons) would have its greatest explanatory power.

References

Algaze, G., R. Breuninger, C. Lightfoot, and M. Rosenberg. 1991. The Tigris-Euphrates Archaeological Reconnaissance Project: A Preliminary Report of the 1989–1990 Seasons. *Anatolica* 17:175–240.

Bates, D.G., and S.H. Lees. 1979. The Myth of Population Regulation. In *Evolutionary Biology and Human Social Behavior: An Anthropological Perspective*, ed. N.A. Chagnon and W. Irons, pp. 273–289. Duxbury Press, North Scituate, MA.

Benedict, P. 1980. Survey Work in Southeastern Anatolia. In *Prehistoric Research in Southeastern Anatolia*, Vol. I, ed. H. Çambel and R.J. Braidwood, pp. 151–191. Istanbul University, Faculty of Letters Press. Istanbul.

Binford, L.R. 1968. Post Pleistocene Adaptations. In *New Perspectives on Archaeology*, ed. L.R. Binford and S.R. Binford, pp. 313–341. Aldine, Chicago.

Blanton, R.E., and J. Taylor. 1995. Patterns of Exchange and Social Production of Pigs in Highland New Guinea: Their Relevance to Questions About the Origins and Evolution of Agriculture. *Journal of Archaeological Research* 3:113–145.

Boyd, R., and P.J. Richerson. 1985. *Culture and the Evolutionary Process*. University of Chicago Press, Chicago.

Bronson, B. 1975. The Earliest Farming: Demography as Cause or Consequence. In *Population, Ecology, and Social Evolution*, ed. S. Polgar, pp. 53–78. Mouton, The Hague.

Byrd, B.F. 1994. Public and Private, Domestic and Corporate: The Emergence of the Southwest Asian Village. *American Antiquity* 59:639–666.

Chagnon, N.A. 1983. *Yanomamo: The Fierce People*. 3d ed. Holt, Rinehart and Winston, New York.

Charles, D.K. 1992. Shading the Past: Models in Archaeology. *American Anthropologist* 94:905–925.

Childe, V.G. 1951. *Man Makes Himself*. Mentor Books, New York.

Clark, G.R. 1979. Seasonal Growth Variations in the Shells of Recent and Prehistoric Specimens of *Mercenearia* from St. Catherine's Island, Georgia. *Anthropological Papers of the American Museum of Natural History* 56:161–179.

Cowan C.W., and P.J. Watson (eds.). 1992. *The Origins of Agriculture: An International Perspective*. Smithsonian Institution Press, Washington.

Cowgill, G.L. 1975. On the Causes and Consequences of Ancient and Modern Population Changes. *American Anthropologist* 77:505–525.

Dawkins, R. 1989. *The Selfish Gene*. 2d ed. Oxford University Press. Oxford.

Ellen, R. 1982. *Environment, Subsistence and System: The Ecology of Small-Scale Social Formations*. Cambridge University Press, Cambridge.

Flannery, K.V. 1969. Origins and Ecological Effects of Domestication in Iran and the Near East. In *The Domestication and Exploitation of Plants and Animals*, ed. P.J. Ucko and G.W. Dimbleby, pp. 73–100. Aldine, Chicago.

——— 1973. The Origins of Agriculture. *Annual Review of Anthropology* 2:271–310.

1982. Early Pig Domestication in the Fertile Crescent: A Retrospective Look. In *The Hilly Flanks and Beyond: Essays on the Prehistory of Southwestern Asia*, ed. T. Cuyler-Young, Jr., P.E.L. Smith, and P. Mortensen, pp. 163–188. Studies in Ancient Oriental Civilization 36, University of Chicago, Oriental Institute, Chicago.

Harris, D.R. 1978. Settling Down: An Evolutionary Model

for the Transformation of Mobile Bands into Sedentary Communities. In *The Evolution of Social Systems*. ed. J. Friedman and M. Rowlands, pp. 401–417. University of Pittsburgh Press, Pittsburgh.

Hayden, B. 1990. Nimrods, Piscators, Pluckers and Planters: The Emergence of Food Production. *Journal of Anthropological Archaeology* 9:31–69.

——— 1992. Models of Domestication. In *Transitions to Agriculture in Prehistory*, ed. B. Gebauer and T.D. Price, pp. 11–19. Prehistory Press, Madison.

——— 1995. A New Overview of Domestication. In *Last Hunters–First Farmers*, ed. T.D. Price and A.B. Gebauer, pp. 273–299. School of American Research Press, Santa Fe.

——— 1996. Feasting in Prehistoric and Traditional Societies. In *Food and the Status Quest*, ed. P. Wiessner and W. Schiefenhovel, pp. 127–147. Berghahn Books, Providence.

Henry, D.O. 1989. *From Foraging to Agriculture: The Levant at the End of the Ice Age*. University of Pennsylvania Press, Philadelphia.

1991. Foraging, Sedentism, and Adaptive Vigor in the Natufian: Rethinking the Linkages. In *Perspectives on the Past: Theoretical Biases in Mediterranean Hunter-Gatherer Research*, ed. G.A. Clark, pp. 353–370. University of Pennsylvania Press, Philadelphia.

Higham, C., and B. Maloney. 1989. Coastal Adaptation, Sedentism, and Domestication: A Model for Socio-Economic Intensification in Prehistoric Southeast Asia. In *Farming and Foraging: The Evolution of Plant Exploitation*, ed. D.R. Harris and G.C. Hillman, pp. 650–666. Unwin Hyman, London.

Hole, F. 1989. A Two-Part, Two-Stage Model of Domestication. In *The Walking Larder: Patterns of Domestication, Pastoralism, and Predation*, ed. J. Clutton-Brock, pp. 97–104. Unwin Hyman, London.

Jochim, M.A. 1976. *Hunter-Gatherer Subsistence and Settlement: A Predictive Model*. Academic Press, New York.

Keeley, L.H. 1988. Hunter-Gatherer Economic Complexity and 'Population Pressure': A Cross-Cultural Analysis. *Journal of Anthropological Archaeology* 7:373–411.

Lawrence, B. 1982. Principal Food Animals at Çayönü. In *Prehistoric Village Archaeology in South-Eastern Turkey*, ed. L.S. Braidwood and R.J. Braidwood, pp. 175–199. BAR International Series 138. Oxford.

Lee, R.B. 1969. Eating Christmas in the Kalahari. *Natural History* 78:14–22, 60–63.

Lee, R.B., and I. DeVore. 1968. Problems in the Study of Hunter-Gatherers. In *Man the Hunter*, ed. R.B. Lee and I. DeVore, pp. 3–12. Aldine, Chicago.

MacNeish, R.S. 1992. *The Origins of Agriculture and Settled Life*. University of Oklahoma Press, Norman.

McCorriston, J., and F. Hole. 1991. The Ecology of Seasonal Stress and the Origins of Agriculture in the Near East. *American Anthropologist* 93:46–69.

Maynard Smith, J. 1982. *Evolution and the Theory of Games*. Cambridge University Press, Cambridge.

Meyers, J.T. 1971. The Origins of Agriculture: An Evaluation of Three Hypotheses. In *Prehistoric Agriculture*, ed. S. Streuver, pp. 101–121. The Natural History Press, New York.

Miller, N.F. 1996. Seed Eaters of the Ancient Near East: Human or Herbivore? *Current Anthropology* 37: 521–528.

Olszewski, D.I. 1993. Subsistence Ecology in the Mediterranean Forest: Implications for the Origins of Cultivation in the Epipaleolithic Southern Levant. *American Anthropologist* 95:420–435.

Price, T.D., and J.A. Brown. 1985. Aspects of Hunter-Gatherer Complexity. In *Prehistoric Hunter-Gatherers: The Emergence of Cultural Complexity*, ed. T.D. Price and J.A. Brown, pp. 3–20. Academic Press, Orlando.

Quitmeyer, I.R., S. Hale, and D.S. Jones. 1985. Paleoseasonality Determination Based on Incremental Shell Growth in the Hard Clam, *Mercenaria mercenaria*, and Its Implications for the Analysis of Three Southeast Georgia Shell Middens. *Southeastern Archaeology* 4(1): 27–40.

Redding, R.W. 1994. The Origin of Food Production in Anatolia: A Test of 'Foreign' Ideas. Paper presented at the annual meeting of the Society for American Archaeology, April 20–24, 1994. Anaheim, CA.

Redding, R.W. 1988. A General Explanation of Subsistence Change: From Hunting and Gathering to Food Production. *Journal of Anthropological Archaeology* 7:59–97.

Rindos, D. 1984. *The Origin of Agricultural Systems: An Evolutionary Perspective*. Academic Press, New York.

Rosenberg, M. 1990. The Mother of Invention: Evolutionary Theory, Territoriality, and the Origins of Agriculture. *American Anthropologist* 92:399–415.

——— 1994a. Hallan Çemi Tepesi: Some Further Observations Concerning Stratigraphy and Material Culture. *Anatolica* 20:121–140.

——— 1994b. Pattern, Process, and Hierarchy in the Evolution of Culture. *Journal of Anthropological Archaeology* 13:307–340.

——— n.d. Cheating at Musical Chairs: Sedentism in an Evolutionary Context. *Current Anthropology*. In press.

Rosenberg, M., and M. Davis. 1992. Hallan Çemi Tepesi,

an Early Aceramic Neolithic Site in Eastern Anatolia: Some Preliminary Observations Concerning Material Culture. *Anatolica* 18:1–18.

Rosenberg, M, R.M.A. Nesbitt, R.W. Redding, and T.F. Strasser. 1995. Hallan Çemi Tepesi: Some Preliminary Observations Concerning Early Neolithic Subsistence Behaviors in Eastern Anatolia. *Anatolica* 21:1–12.

Rosenberg, M., and R.W. Redding. 1995. Hallan Çemi, Feasting, and the Beginnings of Settled Village Life in Highland Southwestern Asia. Paper presented at the annual meeting of the Society for American Archaeology, Minneapolis, May 3–7, 1995.

Rosenberg, M., R.W. Redding, and A. Akin. n.d. Hallan Çemi and Early Village Organization in Eastern Anatolia. In *Social Configurations of the Near Eastern Early Neolithic: Community Identity, Heterarchical Organization, and Ritual*, ed. I. Kuijt. In press.

Rosenberg, M., and H. Togul. 1991. The Batman River Archaeological Site Survey, 1990. *Anatolica* 17: 241–254.

Steward, J.H. 1938. *Basin-Plateau Aboriginal Sociopolitical Groups*. Bureau of American Ethnology Bulletin 120. Washington, DC.

Wills, W.H. 1988. *Early Prehistoric Agriculture in the American Southwest*. SAR Press, Santa Fe.

Wright, G.A. 1971. Origins of Food Production in Southwestern Asia: A Survey of Ideas. *Current Anthropology* 12:447–477.

Yellen, J. 1985. Bushmen. *Science* 85:41–48.

——— 1990. The Transformation of the Kalahari !Kung. *Scientific American* 262(4): 96–105.

ANCESTRAL PIGS: A NEW (GUINEA) MODEL FOR PIG DOMESTICATION IN THE MIDDLE EAST[1]

Richard W. Redding

Museum of Anthropology, 1109 Geddes Avenue, University of Michigan, Ann Arbor, MI 48109-1079

Michael Rosenberg

University Parallel Program, University of Delaware, 333 Shipley St., Wilmington DE 19801

Introduction

The Middle East has long been recognized as an area in which food production developed at an early date, and a substantial body of data on the origins of food production in that area has been and is available. These data have produced a consensus concerning the domestication of animals in the Middle East: that the earliest domesticates were sheep and goats (e.g., Hole 1984, 1989). Sheep are thought to have been domesticated before goats, and these two taxa are thought to have been followed later by pigs and cattle. Generally, pigs have been cast in the role of a late and rather unimportant addition to the repertoire of Middle Eastern animal domesticates despite the presence of pigs, sometimes in large numbers, at early food producing sites such as Çayönü (Lawrence 1980, 1982), Jarmo (Reed 1960; Flannery 1982), Beidha (Hecker 1975), Jericho (Clutton-Brock 1979), Asiab (Bökönyi 1977), Ali Kosh (Flannery 1969), and others. We would suggest two reasons for the consensus view that pigs are a late, minor addition to the repertoire of domesticates in the Middle East. The first is that much of the data on early pig domestication are from sites that were probably peripheral to areas in which pigs might have been domesticated. Flannery (1982:182) recognized that Jarmo, Ali Kosh, and other sites along the Zagros Mountains were on the periphery of pig domestication and suggested that we should look at sites in southeast Europe, the Zagros Mountains, and the southern Taurus for the earliest evidence of pig domestication. We will argue that a second reason for the consensus view is that ideas and models of domestication seem to be rooted in the strategies and tactics of animal use employed in the Middle East at present and those that were prevalent in Europe during later periods. In both of these areas sheep, goats, and cattle dominate the domestic fauna, and the strategies and tactics used in herding these animals form, for many researchers, their concept of what is domestic and how to identify domestication. While this is not made explicit in discussions of early animal domestication in the Middle East, it is implicit in most of them. Recent work at the early village site of Hallan Çemi in eastern Turkey (Rosenberg and Davis 1992; Rosenberg 1994; Rosenberg et al. 1995) indicates that there are serious errors in this consensus view, particularly as regards the role and timing of pig use in the shift from hunting-gathering to food production.

This paper will briefly explain and critique the methods whereby the status of pigs and other early domesticates has been determined in archaeologically derived faunal samples. This will be followed by a brief description of pig use in New Guinea. The New Guinea data, along with the Hallan Çemi[2] data, will then serve as the basis for developing an alternative model of the role of pigs in the earliest stages of the shift to food production in the Middle East.

Domestic or Wild?

The basic problem confronting archaeologists working with faunal remains from sites associated with the shift from hunting-gathering to food production is determining how the animal taxa were used by the site's inhabitants. This problem is usually framed in a discussion of whether the various taxa were domestic or wild. Traditionally, the state of one or more of six criteria has been used to argue the status of an individual taxon. These include change in morphology, body or body part size, species composition, age structure, sex ratio, body part distribution, and changes in relative abundance of taxa. Other criteria have been used (e.g., pathologies, artistic representations, paraphernalia for controlling domesticated animals), but it is the six listed above and discussed below that are most frequently used and on which we will focus.

Two of the most commonly employed criteria, morphology and body or body part size, are related and are sometimes difficult to separate. For sheep and goats, morphological changes in horn core shape and size have been used (e.g., Flannery 1969; Stampfli 1983). In size change, studies of measurements on one or more elements of a taxon's skeleton are taken and compared with similar measurements on samples drawn from known wild and domestic taxa. Change in size, particularly a reduction in size, has been used to indicate domestication in sheep and goats (e.g., Uerpmann 1979; Helmer 1989), cattle (e.g., Grigson 1989), and pigs (Flannery 1982). The third and fourth criteria are sometimes referred to as demographic evidence (Meadow 1989). These are the age structure and the sex ratio of the population of each taxa consumed based on the archaeological sample. A disproportionate presence or a deviation from an expected value in an archaeological sample of individuals falling within certain age and sex categories can be construed to indicate culling or other forms of human management (e.g., Perkins 1964, 1973; Hesse 1978). The fifth criterion is body part distribution. To our knowledge this has not been used previously as an indicator of domestication, but has been used to suggest the presence of tactics of animal use in hunting or other, alternative animal use strategies (e.g., Perkins and Daly 1968; Binford 1984; Klein 1989). However, body part distribution provides insight as to whether a taxon is being butchered at or away from the site and, hence, in combination with other criteria may help us determine the taxon's status (an example of this approach will be presented in this paper). The final criterion is relative species abundance. An increase in the representation of a taxon or set of taxa at a site may indicate that the taxon in question is now being maintained and managed (e.g., Bar-Yosef 1981; Payne 1975; Meadow 1984).

As is almost to be expected, problems exist with each of these criteria when taken individually. For example, many researchers have suggested that morphological changes only occur after domestication has already occurred and that the earliest stages of the process may best be evident in other variables (e.g., Bökönyi 1976).

A recent work by Zeder, Cleghorn, and Lapham (1995) has called into question the reliability of size reduction as a marker of early domestication in sheep and goats. Jarman and Wilkinson (1972), Hesse (1984), and Meadow (1989) have raised questions concerning the significance of certain age structures and sex ratios as necessarily being indicative of culling. Species abundance must be used with caution in light of the suggestion that in the earliest stages of domestication a taxon was probably used for "subsistence insurance" and, although domestic, may have formed a minor part of the diet (Redding 1988). Clearly, no single criterion is completely reliable. We suggest that, ideally, the state of all criteria should be considered as a set in an attempt to determine a taxon's status.

For pigs, like cattle, metrics have tended to be the primary method for determining whether the remains at a given site are wild, domestic, or some mixture of both. Their status has usually been determined by measurements of second and third molars (Flannery 1969, 1982); sites yielding pig teeth primarily in the range for domestic animals are assumed to have had domestic pigs, while sites yielding teeth primarily in the range for wild pigs are assumed to have had wild ones. To our knowledge, age structure, sex ratio, body part distribution, and species abundance have not been widely employed to suggest the status of pigs in archaeological samples and have certainly not been used together. However, what if a sample of pig remains was from a site whose inhabitants practiced both hunting and herding during the transition from the former to the latter, or employed tactics of pig husbandry that allowed breeding with wild pigs? In such cases, the expected reduction in tooth size would at best be obscured, even though some pigs were being kept at the site as husbanded resources. In such cases, the other criteria would be crucial to understanding the true status of pigs at that site.

The Tactics of Pig Use in New Guinea

In New Guinea pigs are the most important domestic animal. Further, it is an area in which strategies and tactics of pig use have developed that are quite different from those of Europe and the West in general. As such, New Guinea provides us with alternative models for pig use that can be compared with strategies and tactics used in the earliest Neolithic. While one might object to the use of the New Guinea data as an "analogy," that is not what we intend to do. We will use the strategies and tactics of New Guinea to create a model for pig use in the Middle East and then test that model with data from the Middle East.

Anthropologists studying New Guinean societies that use pigs have long had difficulty with the terms "domestic" and "wild" and have felt compelled to develop other terms that blur the boundary between domestic and wild (see Dwyer 1996). Part of the problem seems to stem from the different aspects of pig use on which researchers focus. For example, several authors have used the term "semi-domestication": e.g., Townsend 1969; Hughes 1970; Boyd 1984; Morren 1986; Kelly 1988; Yen 1991. For Morren, according to Dwyer (1996), semi-domestic pigs are pigs born to domestic mothers and wild fathers; for Kelly they are pigs allowed to roam freely; for Townsend, Hughes, and Boyd, they are pigs maintained in low numbers with little labor invested; and for Yen, semi-domestication is a breeding tactic in which males born in captivity are all castrated and females are mated intentionally to wild males. Clearly some researchers studying pig use in New Guinea are focusing on breeding tactics and others on management tactics.

Dwyer (1996) focuses his analysis of pig use in New Guinea on reproductive status. He accepts a distinction between wild and domestic based on a simple ecologic criterion: wild pigs roam freely, independent of humans, while domestic pigs receive care from humans (ibid.:483). Dwyer readily admits that in New Guinea even this distinction does not allow complete separation of existing pig populations. He accepts a classification system with four possible states that he synthesizes from the work of Baldwin (1990) and Yen (1991), while refining the definitions and terminology. Dwyer does not specifically offer a definition of "wild," but it is implicit in his subsequent definitions that wild pigs are allowed to roam freely and are not interfered with by humans. He defines three categories for pigs maintained by humans based on reproductive status: reproductive alienation, female breeding, and male-female breeding. In reproductive alienation, all captive pigs (pigs which humans are managing) are the result of matings between wild boars and wild sows. The managed population is entirely alienated from breeding, and genetic changes can only result from preferentially selecting animals from the wild population. In female breeding, all managed pigs are the result of matings between wild boars and domestic sows. The managed population consists of females and their young, and genetic changes are the result of selection of females for breeding stock. In male and female breeding, all managed pigs are the result of matings between domestic boars and domestic sows.

Clearly these categories are not mutually exclusive and it is possible for groups to use one, two, or all three of the breeding tactics in various proportions. However, in his analysis of the ethnographic examples from New Guinea, Dwyer (1996:491) finds that while the grade between reproductive alienation and female breeding is well represented, the shift from either reproductive alienation or female breeding to male and female breeding is abrupt, with only one possible intermediate case identifiable in the ethnographic literature.

The use of reproductive alienation or female breeding as a tactic in pig use has advantages. The most obvious is that the human herders avoid the problem of keeping large and typically aggressive males (Meggitt 1958; Dornstreich 1973; Malynicz 1970; Strathern 1984). Also, it requires minimal labor to maintain the husbanded pigs, as they are allowed to roam (and feed) freely. It does, however, require that there be a local population of wild pigs. It also requires either that the pigs and humans not compete for the same food resources or that there be sufficient resources to support both populations.

A Model for Early Pig Use in the Middle East

The use of either or both the reproductive alienation and female breeding tactics in pig exploitation, as developed and described by Dwyer (1996), provides the basis for an alternative to the strategy of domestication developed from Middle Eastern and European models, and applied to pigs in the Middle East. The model we will develop will be based on the use of female breeding. We will ignore reproductive alienation at this time. The use of female breeding in a model presents problems for archaeologists working with the origins of food production. First, how does one identify, based on the faunal remains from a site, the results of a strategy based on female breeding—can we develop a predictive model? Second, how does such a strategy fit in with the theory and explanation for the origin of food production?

It is now widely recognized that for hunter-gatherers the abandonment of mobile lifeways in favor of sedentary ones carries with it the very real risk of local resource depletion (e.g., Hayden 1981; Binford 1983; Brown 1985; Lieberman 1993). One of us (Redding 1988) has argued that risk reduction, through subsistence insurance, was the major selective force operating in the shift from hunting-gathering to food production. We argue here that in areas inhabited by wild pigs, such as the ancient Middle East, female breeding of pigs would provide a readily available source of energy, fat, and protein that could have functioned as a very effective, low-cost form of subsistence insurance. The inhabitants of the earliest sedentary societies in this region could easily have obtained young female pigs to rear within the settlement. These female pigs could have roamed freely in and around the settlement and been allowed to mate with wild males. The resulting female offspring would have been either kept for further breeding or consumed; the male offspring would have been freely consumed. These young pigs would have been a source of energy, fat, and protein. Moreover, if the early function of maintaining pigs was subsistence insurance, then they need not have been kept in large numbers. That is, they were essentially a reserve, to be used primarily when other sources of energy, fat, or protein were not readily available for whatever reason.

If we ignore plant resources for the moment and consider only the animals potentially capable of being domesticated, the pig is the animal best suited to function as subsistence insurance. This is the case for at least five reasons:

1. The fecundity and growth rate of pigs makes them the superior producer of energy, fat, and protein relative to all other domestic animals except the chicken (see Redding 1991). Harris (1985) states that pigs convert 35% of the energy in their feed to meat, compared to 13% for sheep and a mere 6.5% for cattle. Data presented by Pimental and Pimental (1979:55–61) indicate that pigs convert 16.2% of the energy they consume to protein, feedlot cattle convert 12.6%, and free range cattle convert only 1%. The discrepancies between the two data sets may reflect the fact that Pimental and Pimental only calculate protein yields, ignoring fat. What these data indicate is that pigs turn more of the food they consume into usable meat than do cattle, sheep, or goats.

2. Enhancing the importance of the above is the fecundity of pigs. Pigs normally have litters of 6–10 young and may breed more than once a year. In unimproved breeding of pigs in Nigeria, the average number of pigs reared per sow per year is 14–15 (Williamson and Payne 1978:554). Cattle normally rear less than 1 calf per year and sheep in the Middle East rear about 1.2 lambs per year (Redding 1981).

3. Given the simple husbandry system envisioned, the labor required for pig maintenance is lower than for sheep, goats, or cattle. Unless specific reasons exist to do otherwise, they need not be herded and may be left to forage in and around the village with very little supervision. The tactic of female breeding would further reduce the labor that need be invested.

4. Young pigs tame readily and, like dogs, will imprint on humans.

5. Juvenile or neonate pigs are relatively easy to obtain. Females create nests and leave their young in them during their crepuscular feeding forays. The young remain in these nests for several weeks. If nests are known, then it is relatively easy to obtain young pigs. In 1976, one of us (Redding) obtained two piglets in this fashion while in Iran. Also, Charles Reed (1960:139) notes that he kept two during excavations at Jarmo.

This female breeding model generally requires that three conditions be met in order for it to become operational and all three have been alluded to above. The first is that a population of wild pigs be available near the settlement. The second is that the human population be sedentary. Pigs are difficult to herd and pig husbandry of any sort would thus be incompatible with a mobile way of life. The third relates to the plant resources being exploited by the human population, and competition with pigs for resources. If the human population were engaged in the intensive exploitation of cereals (wild or domestic), then female breeding would likely be an impractical option. This is because wild pigs compete directly with human for cereals; it would take considerable effort to consistently keep wild pigs away from local stands of grain while maintaining a sufficient population of wild boars for breeding. It is very likely that the extra labor required to keep wild pigs and domesticated females away from the stands or fields of cereals would negate the above-enumerated positive attributes of pigs and rule out female breeding as a practical option. Hence, when a human group shifts toward use of wild cereals or cultivation of cereals, they would drop female breeding of pigs. They would either shift tactics of pig use, possibly to male and female breeding, or drop pigs from their subsistence system.

Female Breeding and the Archaeological Record

If the inhabitants of a hypothetical site were exploiting pigs through a system of female breeding, how can we recognize this using the faunal remains from that site? Clearly, under a system of female breeding, any reduction in molar tooth size resulting from human selection of female breeding stock would be mitigated or lost by allowing the uncontrolled breeding of these females with wild males. Hence, to test for the occurrence of this tactic of pig use at a site, we must look to criteria beyond just tooth size reduction.

Fortunately, if a strategy based on female breeding were being employed, it should produce a pattern in three other variables. First, the population of pigs consumed at the site should resemble consumption from a domestic population in terms of age structure as reflected in survivorship. Specifically, the heavy use of young pigs should be evident in survivorship data indicating that 40–60% of the pigs consumed were under one year of age (e.g., Redding 1991). Second, a sex ratio biased toward males should obtain, as females are preferentially being retained for breeding while males are preferentially being consumed. Third, the ratio of body parts at the site should reflect the butchering of animals at the site. Specifically, the ratio of meat-bearing to non-meat-bearing limb fragments should be similar to their relative frequency in the pig skeleton.

The final criterion available to us, changes in species abundance, is problematical. We might expect the proportion of pig remains in the sample to increase with a shift toward female breeding. However, as already noted, this is not a necessary outcome of such a shift. If the tactic of female breeding is used to create an insurance resource, then pigs may be used in about the same proportions as they were when used as a wild resource. If an increase in the proportion of pigs at a site is identified, this would suggest a shift in the system of pig use and, in conjunction with predicted states in other variables, would be indicative of female breeding. But an increase is not a necessary condition of the shift.

The Pig Remains from Hallan Çemi

Hallan Çemi is an aceramic site in the Taurus foothills of southeastern Turkey that was occupied over the course of several hundred years at the end of the eleventh millennium BP (uncalibrated). It represents the remains of a sedentary society that subsisted primarily by means of hunting and gathering (see Rosenberg 1994; Rosenberg et al. 1995; Rosenberg and Redding, this volume). The most intensively exploited plant resources were nuts and pulses (cereals are absent) and the most intensively exploited animals were sheep and deer.

Hallan Çemi would have been well suited for the practice of maintaining pigs by means of female breeding. First, it is located above a small river, the Sason Çayı, which would have provided lush vegetation and bedding sites for wild pigs. Further, the surrounding hills (even now largely covered by oak-pistachio forests) would have provided ample forage for a population of pigs. Second, there is no evidence for the use of grains by the site's inhabitants. Hence, humans and wild pigs would not have been competing directly for this potential food resource. It is true that both humans and pigs would have been feeding on the nut crops. However, the site's inhabitants seem to have concentrated their energies on the exploitation of almonds and pistachios (see Rosenberg et al. 1995). This presumably left the bulk of the more plentiful acorns unused and thus available to pigs. The point is that the site's inhabitants are unlikely to have seen pig competition for nuts as having a negative impact on their subsistence. Finally, the evidence (see Rosenberg and Redding in this volume) indicates that Hallan Çemi was occupied year-round. Thus, the above-mentioned three conditions required for successful female breeding of pigs are met at Hallan Çemi.

What follows is based on a sample of over 35,000 pieces of bone analyzed to date. Of these, ca. 4,300 were limb, skull, and teeth fragments that could be identified. Included within this total are over 845 fragments of pig. Thus far, only about 20% of the material to be studied has been analyzed, so the data set will expand as more work is done. The material that has been examined to date comes from two main contexts. The first is a central "pit" that appears to have been a wind-deflated central area that was, among other things, used for trash disposal. Excavations in this area have reached a depth of more than 4 m. The second context is in and around a series of elaborate structures that surround the central area.

In general, the faunal remains from the site are in extremely good condition. Evidence of carnivore gnawing is rare, and soft element ends (e.g., proximal humerus, distal femur, proximal tibia) are present in both fused and unfused states. All this suggests that the sample has not been heavily biased by non-human activities and that any patterning in the sample is probably the result of human subsistence behavior.

The dominant mammals are deer (NISP[3]=1137), sheep and goats (NISP=1689), and pig (NISP=845). The deer material is primarily from the red deer (*Cervus elaphus*), although the fallow deer (*Dama dama*) is present at low levels. Limb fragments for which fusion could be determined make up 371 of the deer total. The age profile based on these fragments indicates that 93% survived to 10 months, 90% to 16 months, 66% to 24 months, 53% to 36 months, and 50% to 42 months. The ratio of fragments from non-meat-bearing limb bones to fragments from meat-bearing limb bones is 1.87. The ratio expected if whole animals were being butchered on the site is 0.5.

The sheep and goat material exhibits no morphological evidence of domestication. The horn cores are all of the wild phenotype. The sheep and goat material can be mostly attributed to wild sheep (*Ovis ammon*), as the ratio of identifiable sheep to identifiable goat is 25:1. Limb fragments for which fusion could be determined make up 521 of the sheep and goat total. The age profile based on these fragments indicates that 96% survived to 10 months, 81% to 16 months, 70% to 24 months, 69% to 36 months, and 68% to 42 months. The ratio of fragments from non-meat-bearing limb bones to fragments from meat-bearing limb bones is 2.32. The ratio expected if whole animals were being butchered on the site is 0.5.

For the purposes of this paper it is the material representing the pig that is critical. We need to examine the data derived from the pig material in detail, beginning with measurements made on the teeth and proceeding to the data for each of the other criteria.

Measurements for 24 upper or lower, second or third molars from Hallan Çemi are provided in Table 1, along with summary statistics. A comparison with the data assembled by Flannery (1982) is in order. The Hallan Çemi lower third molars have a lower mean length (40.5

Table 1. Measurements on teeth of *Sus scrofa* from Hallan Çemi

Tooth		Length (mm)	Breadth (mm)
Upper second molar		26.3	–
		25.3	–
		24.9	–
		25.5	–
		22.1	–
		22.4	20.3
		24.1	20.7
		21.1	19.0
		21.8	20.3
		19.6	22.2
	Mean	23.3	20.5
	Std. dev.	2.213	–
Upper third molar		39.4	23.9
		38.3	–
		37.7	–
		35.3	–
		36.8	–
		35.5	–
	Mean	37.2	–
	Std. dev.	1.61	–
Lower second molar		23.2	16.3
		22.0	16.4
		19.6	15.3
	Mean	21.6	16.0
Lower third molar		40.0	19.0
		41.9	22.0
		38.4	17.7
		41.3	17.6
		40.9	18.9
	Mean	40.5	19.04
	Std. dev.	1.362	–

the Hallan Çemi sample are larger except for the lower second molar and this is probably a function of sample size at Hallan Çemi (see Table 1).

The samples from the pottery-bearing levels of Jarmo were large enough to run a t-test comparing them to the Hallan Çemi samples for the upper M3, upper M2, and the lower M3. The tests on the upper M3 and lower M3 indicated that the differences are significant (upper M3 - p=0.0129, 5df: lower M3 - p=0.0351, 2df). The test on the upper M2 was very nearly significant (p=0.0502, 9df). A comparison of the Hallan Çemi teeth measurements to those provided by Flannery (1982:169) for modern wild pigs from southwest Asia is of interest. The mean length for the upper M3 in 21 recent skulls is 37.4, while the mean for the Hallan Çemi sample is 37.2. The mean length for the upper M2 in 21 recent skulls is 23.3, while the mean for the Hallan Çemi sample is 23.3. Finally, the mean for the lower M3 for the recent 21 skulls is 41.3, while for the Hallan Çemi sample it is 40.5. Not surprisingly, t-tests on the upper M3, upper M2, and lower M3 all showed no significant difference between the two samples (upper M3 - p=0.716, 5df: upper M2 - p=6315, 9df: lower M3 - p=0.6052, 4df). Finally, I (Redding) used summary statistics (standard deviation and sample size) provided by Kuşatman (1991) for length measurements of upper third, upper second, and lower third molars for a sample of modern wild pigs from Turkey to perform three t-tests. All three tests indicated no significant differences (p>0.25) between the two samples. Two caveats are in order at this point. First, the sample of measured teeth is small and will grow as more of the Hallan Çemi sample is done. Second, populations of wild pigs in the Middle East undoubtedly exhibit (and have exhibited) geographic variation.[4] Hence, when comparing measurements from sites we must realize we are dealing with both temporal and geographic variation. What is clear is that, at this time, teeth measurements do not provide evidence that the pigs at Hallan Çemi were domestic. In fact, the teeth measurements suggest that the population of pigs being used by the inhabitants of Hallan Çemi was wild.

versus 42.6) than the sample of presumably wild pigs from Karim Shahir and Tepe Asiab (Flannery 1982:171). However, a t-test of these data suggests that no significant difference exists between the two samples (p=0.44, 2df). While Flannery does not supply summary statistics for the sample of pig teeth from the aceramic levels of Jarmo, probably because of small sample size, the Hallan Çemi teeth appear to be slightly smaller based on the measurements Flannery provides (ibid.:173). Flannery does provide summary statistics for the pottery-bearing levels of Jarmo (ibid.:175), and the teeth in

At present the best data set for age structure for the pigs consumed at Hallan Çemi is from fusion of limb elements. The sample of limb fragments that provide fusion data for pigs at Hallan Çemi is 238; the following analysis uses 125 of these fragments. In pigs, the supraglenoid tuberosity of the scapula, the distal humerus, and the proximal epiphysis of the second phalanx all fuse at about one year, while the proximal and distal ulna, distal radius, proximal humerus, and distal femur fuse between three and three-and-one-half years. At Hallan Çemi, the fusion data for these elements indicate that about 40% of the pigs at the site were killed before they reached one year of age and that about 46% were killed after they attained the age of three years. The age structure of the pigs killed and consumed at Hallan Çemi is quite different from the age structures for sheep-goats and red deer remains, both of which are almost certainly the result of hunting. On the other hand, the Hallan Çemi pattern of pig consumption is very similar to those found at later Middle Eastern sites at which clearly domestic pigs were kept (e.g., Girikihaciyan; see McArdle 1990:110).[5]

The sex can be determined for pig canines and for pubis fragments: the former using size and the latter using the shape of the ilio-pectineal eminence. Normally we prefer to use the sex ratio derived from the two sources as a check on each other, but at present the samples are too small. To date 15 pig fragments could be sexed, including 9 canines and 6 pubis fragments. Eleven of these were from males and only 4 from females. While this sample too is still small, it is strongly male biased. A strong male bias would be congruent with a model of female breeding and anomalous for a hunted population.

Lastly, at Hallan Çemi the body part distribution data suggest that more of the pigs were being butchered at the site than were red deer or sheep-goats. Specifically, if complete skeletons of pigs were being introduced into the Hallan Çemi deposits, then the ratio of proximal, meat-bearing bones to distal, non-meat-bearing bones should be 0.40. Based on the pig material identified to date, the ratio is 1.02. This is higher than expected, but still much lower than the ratios for both red deer and sheep-goat (see above: red deer 1.87 and sheep-goat 2.32), both of which are presumed to have been exclusively hunted and thus, typically, butchered off-site.

The data from pigs for the criteria used to determine domestication present an interesting picture. The metrics on the teeth indicate that wild pigs are being used. The age structure and sex ratio data indicate that female breeding may have used as a tactic in incipient domestication. The body-part distribution data seem to indicate that a much larger percentage of pigs were being butchered on site than were red deer and sheep. Two points need to be made. First, it should be noted that the beginning of domesticatory practices centered on a particular species does not in itself require that the hunting of wild members of that species suddenly stop. Thus, the above data, indicating less than total on-site butchering of pigs, are not necessarily inconsistent with the earliest stages of pig domestication. The same can be said for the presence of pig teeth from clearly wild individuals. While both the body-part and tooth-size data suggest that some hunting of pigs was ongoing, other aspects of the body-part data coupled with the sex ratio and survivorship data suggest some degree of pig husbandry at Hallan Çemi. The new evidence from nearby Çayönü (see Hongo and Meadow, this volume), suggesting some degree of pig husbandry in even the earliest[6] levels of that site, is also consistent with the early husbandry of pigs at Hallan Çemi. That is, it apparently documents another relatively early example of the practice, thus providing an element of continuity.

Second, the pig data as a whole come from over 4 m of deposits that span several hundred years. It is, therefore, quite possible that during the earliest stages of the site's occupation, the inhabitants obtained pigs primarily, if not completely, by means of hunting, with pig husbandry (female breeding) entering the subsistence equation only toward the later stages. If we limit an examination of the pig material to just one context, the central area (see above), in order to control for biases introduced by disposal patterns, and compare the pig data from the lower 3 m to that for the upper 1 m, some interesting preliminary results obtain. First, let us examine the only criteria we have not used, relative abundance, and its change through time. In the lower 3 m of the central area, deer are represented by 521 fragments, sheep-goat by 591, and pig by 253. Hence, the ratios of deer to pig and sheep-goat to pig are 2.06 and 2.53, respectively. In the upper 1 m of the central area, deer are represented by 191 fragments, sheep-goat by 204, and pig by 187. The ratios of deer to pig and sheep-goat to pig are, respectively, 1.02 and 1.09. The representation of pigs in the upper meter of the central area increases dramatically. The age structure data from the lower 3 m indicate that 83% of the pigs consumed were older than one year (30 of 36 fragments were fused), while in the upper meter of the central area only 50% of the pigs consumed were older than one year (33 of 66 fragments were fused). The body-part distribution data for the lower 3 m of the central area provide a ratio of meat-bearing to non-meat-bearing fragments of 3.0, with a ratio of 0.40 being expected if whole pigs were being butchered on the site. The ratio in the upper meter of the central area is 0.60. This is very close to the expected 0.40. These data suggest that in the upper meter whole pigs are being slaughtered at the site, while in the lower 3 m pigs are being slaughtered off the site. The numbers of measured teeth and sexed fragments are too small to compare between the lower 3 and upper 1 meters. The data on relative abundance, age structure, and body part distribution indicate a dramatic change in the way humans were utilizing pigs between the lower 3 m and the upper 1 m of the central area. A change in the upper meter is strongly congruent with, at the very least, the use of female breeding.

The upper meter of the site demonstrates an increase in objects associated with social stratification (batons, wands, and incised stone bowls). It is in the upper stratum that public buildings clearly appear at the site for the first time (Rosenberg 1994; Rosenberg et al. 1995). This evidence for increased social complexity could conceivably play a role in the shift from hunting pigs to a tactic of female breeding as part of a change in subsistence strategy (cf. Hayden 1990, 1992; see also Rosenberg and Redding, this volume).

Discussion

The totality of the pig data from Hallan Çemi does not support the conclusion that all the animals consumed at that site were wild. It remains unclear whether the pigs being husbanded were specifically being maintained by means of female breeding, but the data for at least the upper meter of the central area is congruent with the use of such a tactic. However, it does not appear that any husbanded animals in the Middle East were domestic in the usual sense.

Let us for the moment assume that, at the earliest stages of the shift to food production, pigs were specifically kept using the tactic of female breeding or some essentially similar non-intensive method. An interesting pair of related questions then arise. The first is why such practices, involving only minimal labor, are ultimately replaced in the region by sheep and goat husbandry, which presumably required more labor. The second is why pig husbandry practices ultimately also change, such that free breeding with wild stock is halted. This was presumably also at the expense of increased labor and resulted in the morphological changes that characterize domestic pigs.

The answers to both are implicit in the model presented here and extensions from it. Specifically, the answers likely lie in the shift from nut to cereal grass exploitation that appears to occur along the upper Tigris at the time of the abandonment of Hallan Çemi and the initial occupation of the region's late round house horizon sites (e.g., basal Çayönü, Nemrik 9, Qermez Dere, M'lefaat).

As noted earlier, pigs are direct competitors with humans for grains and are difficult to control. So, with the shift to intensive cereal grass exploitation, wild pigs would be eliminated from the vicinity of the site as competitors, and the managed pigs could no longer be routinely left to forage unsupervised near the village. Hence, the use of a tactic of female breeding would no longer be viable. Either domestic pigs would be dropped from the subsistence system or the tactic of pig use should shift to male and female breeding. The study by Dwyer (1996) suggests that if a shift to male and female breeding obtains it should be nearly instantaneous. But domestic pigs cannot be easily driven long distances to forage away from the village (and its associated resources), as can sheep and goats. Thus, the model as presented thus far generates the expectation that pig husbandry becomes significantly more labor intensive and thus presumably declines in importance with the shift to the intensive exploitation of cereal grasses. In this event, should animal husbandry still be desired, one could further expect that the lessons initially learned by working with pigs will be applied to more behaviorally tractable species such as sheep and goats. Such animals could be more readily kept away from resources that people want for their exclusive use.

The Hallan Çemi pig data aside, that pigs were domesticated before sheep and goats now seems clear on the basis of the Çayönü data alone (Hongo and Meadow, this volume). That they were also domesticated before the shift to intensive cereal grass exploitation is indicated by the Hallan Çemi data. However, the Cayönü data indicate that pig husbandry continued during the full course of that site's occupation, despite the fact that cereal grasses were also being exploited all that time. In fact, there is clear evidence for continued morphological change (i.e., size reduction) in pigs over the entire Çayönü sequence (Hongo and Meadow, this volume).

Why were pigs retained at Çayönü after the domestication of cereals? The continued size reduction of the pigs at Çayönü implies increased restrictions on mating; in other words, increased human control. Clearly we are seeing the introduction of (shift to?) male and female breeding. Whether the primary purpose of such increased control was to restrict the access of pigs to food resources such as cereals or to restrict mating is beside the point; it likely accomplished both. The question thus becomes, Why control pigs rather than sheep and goats at this stage of the process?

The answer probably lies in the fact that along with a shift in the tactics of pig use, the function of pigs shifted: i.e., the selective value of pig herding shifted. Two components of the selective value of pigs need to be examined. The first has to do with subsistence. It was suggested that the initial impetus to produce food animals was as subsistence insurance against any resource depletion resulting from sedentism. The point is that animal insurance does not require large numbers of animals. In terms of labor, the ultimate advantage of pigs is that they do not require managing unless circumstances dictate otherwise. The ultimate advantage of sheep and goats is that they can be managed in very large numbers with little difficulty. However, in small numbers pigs are arguably only slightly less manageable than equal numbers of sheep and goats, while still maintaining their advantage over sheep and goats in rate of food conversion to meat and in fecundity. For example, the data from New Guinea (e.g., the film *Dead Birds*) suggest that even a child can tend small groups of pigs in the vicinity of a village, if all that "tending" entails is keeping them out of producing gardens. One of us (Redding 1991, 1994) has argued that in villages in the ancient Middle East and Egypt pigs functioned as a family resource in areas where access to sheep and goats was controlled or limited. Pigs may have functioned in a similar way in the early village phases of Çayönü. If substantial amounts of hunted resources were readily available, then pigs might be cheaper for individual families to keep than sheep and goats.

But we must also consider another aspect of animal use. Elsewhere in this volume (Rosenberg and Redding), we argue that the evidence for early pig domestication along the Tigris illustrates how erroneous simple linear causal models for the origins of food production tend to be. While we maintain that the initial impetus to produce animals was for reasons relating directly to subsistence risk, social forces must also be factored into any coherent model, if for no other reason than that most human action takes place in a social context. For example, it is entirely possible that the real advantage that pigs enjoyed over sheep and goats by the late round house horizon, with its shift to cereal exploitation, is their social value for feasting or exchange purposes (cf. Hayden 1990, 1992; Blanton and Taylor 1995). There is evidence for feasting and exchange at Hallan Çemi (Rosenberg et al. 1995), and there is certainly evidence for substantial social complexity at Çayönü, including differential access to prestige goods (cf. Davis n.d.).

There is still also the problem of explaining the eventual shift to sheep and goat herding, and it is here that social considerations are arguably most likely to have played an active role. As noted above, when large herd size is desired and cereal exploitation is an important element of the subsistence equation, then sheep and goats are theoretically the animal domesticates of choice. The implication here is that the desire to intensify animal production was an important consideration underlying the shifting emphasis from pig to ovicaprid husbandry. As also noted above, subsistence insurance does not require large herds. Even in the context of rising human population levels, all that is required is an increased number of small herds, each "tended" by a marginally productive member of the family group to which the herd belongs. The implication here is that the intensification of production beyond insurance levels is solely a product of the desire to generate "wealth," by definition a social phenomenon (e.g., Sahlins 1972). Thus, the implication is that the shift from pig to sheep and goat husbandry was driven by social, not subsistence-related, forces.

Conclusions

The faunal data from early settled village sites along the flanks of the Taurus-Zagros arc have always suggested that pigs were a relatively early domesticate. However, in general, research emphasis had always been on the sheep and goats from these sites. This has been the case for three reasons: (1) because sheep and goats were more numerous; (2) because it was always assumed that the quantitative economic importance of sheep and goats destined them to be the earliest domesticates; and (3) because early researchers in the area assumed, based on the rapid appearance of domestic pigs, that pigs had been domesticated somewhere else (i.e., Anatolia or southeastern Europe).

The most recent data from Hallan Çemi indicate that food production may develop along lines that are quite distinct from those that have been suggested. Sedentism initially seems to develop independently of intensive cereal grass exploitation, and the earliest animal domesticate is the pig. The human subsistence system revolved around the gathering of nuts and pulses, as well as the hunting of sheep, goats, and deer. In addition, pigs were apparently maintained by means of a tactic that we suggest was akin to the female breeding practiced in New Guinea. We further suggest that the initial purpose for such maintenance was as dietary insurance. This tactic was employed until intensive cereal grass exploitation came into practice. All else being equal, we would expect to see a decline in the use of domestic pig and the shift to sheep and goat husbandry with this shift to cereal grass exploitation. The fact that we do not indicates that the function of pig use shifted. This shift is related to changes in the subsistence system and/or the social forces that must also have played a role in the evolution of the developed Middle Eastern Neolithic complex.

At the very least, the faunal remains from Hallan Çemi support the view that pigs were being husbanded, probably using a tactic of female breeding, by the upper meter of the deposits. This places the earliest domestication of pigs in Anatolia, an area in which researchers have predicted it would be found (Flannery 1982:182 and pers. comm. 1997).

Notes

1. The model presented here is considerably more detailed than when originally presented in San Francisco. Much of the detail was added due to the constructive comments and criticisms offered by R.E. Blanton, Richard H. Meadow, K.V. Flannery, and an anonymous reviewer subsequent to the original presentation. Whether or not they are now inclined to view the model more favorably, we think that it is much improved as a consequence of that dialog, and we are grateful for their input.

2. Funding and other support for the Hallan Çemi excavation was provided by the National Geographic Society, the National Science Foundation, Mobil Exploration Mediterranean Inc., the University of Delaware Research Foundation, the University of Delaware, the University of Pennsylvania, the American Research Institute in Turkey, and the University of Michigan.

3. NISP is the Number of Identified Specimens.

4. Kuşatman (1991) in a comprehensive study of a large sample of Old World modern pig skeletons documents the geographic variation in measurements. She found differences in populations from several areas of the Middle East. Of particular interest here is that she found the pig populations of Turkey, Iraq, and Iran to be slightly smaller in size than other pig populations in the Middle East and Europe.

5. A caveat is in order here also. The fact is that we do not know specifically what the age structure data would look like from a clearly hunted population of pigs in the Middle East. We assume that it would appear much more like the data for red deer and wild sheep-goat. Further, we have very few data sets for age structure from domestic populations!

6. The lowermost building level, with its round houses, likely dates to the early or early mid-tenth millennium B.P., uncalibrated (M. Özdoğan, pers. comm. with Rosenberg 1996).

References

Baldwin, J.A. 1982. Pig Rearing and the Domestication Process in New Guinea and the Torres Strait Region. *National Geographic Society Research Reports* 14:31–43.

────── 1990. Muruk, Dok, Pik, Kakaruk: Prehistoric Implications of Geographical Distributions in the Southwest Pacific. In *Pacific Production Systems: Approaches to Economic Prehistory*, ed. D.E. Yen and J. Mummery, pp. 231–257. Australian National Museum, Occasional Papers in Prehistory, 18.

Bar-Yosef, O. 1981. The Epi-Paleolithic Complexes in the Southern Levant. In *Préhistoire du Levant*, pp. 389–408. Colloques Internationaux du CNRS, no. 598. Éditions du CNRS, Paris.

Binford, L.R. 1983. *In Pursuit of the Past: Decoding the Archaeological Record*. Thames and Hudson, London.

────── 1984. *Faunal Remains from Klasies River Mouth*. Academic Press, New York.

Blanton, R.E., and J. Taylor. 1995. Patterns of Exchange and Social Production of Pigs in Highland New Guinea: Their Relevance to Questions About the Origins and Evolution of Agriculture. *Journal of Archaeological Research* 3:113–145.

Bökönyi, S. 1976. Development of Early Stock Rearing in the Near East. *Nature* 264:19–23.

────── 1977. *Animal Remains from Four Sites in the Kermanshah Valley, Iran*. BAR Supplementary Series, 34. Oxford.

Boyd, D.J. 1984. The Production and Management of Pigs: Husbandry Option and Demographic Patterns in an Eastern Highlands Herd. *Oceania* 55:27–49.

Brown, J.A. 1985. Long-Term Trends to Sedentism and the Emergence of Complexity in the American Midwest. In *Prehistoric Hunter-Gatherers: The Emergence of Cultural Complexity*, ed. T.D. Price and J.A. Brown, pp. 201–231. Academic Press, Orlando.

Clutton-Brock J. 1979. The Mammalian Remains from the Jericho Tell. *Proceedings of the Prehistoric Society* 45:135–157.

Davis, M.D. n.d. Social Differentiation at the Early Village of Çayönü, Turkey. Paper presented at the annual meeting of the American Anthropological Association, Chicago, November 20–24, 1991.

Dornstreich, M.D. 1973. An Ecological Study of Gadio Enga (New Guinea) Subsistence. Ph.D. dissertation, Columbia University, New York.

Dwyer, P.D. 1996. Boars, Barrows and Breeders: The Reproductive Status of Domestic Pig Populations in Mainland New Guinea. *Journal of Anthropological Research* 52:481–500.

Flannery, K.V. 1969. The Animal Bones. In *Prehistoric and Human Ecology of the Deh Luran Plain*, ed. F. Hole, K.V. Flannery, and J.A. Neely, pp. 262–330. University of Michigan, Museum of Anthropology, Memoir 1. Ann Arbor.

────── 1982. Early Pig Domestication in the Fertile Crescent: A Retrospective Look. In *The Hilly Flanks and Beyond: Essays on the Prehistory of Southwestern Asia*, ed. T. Cuyler-Young, Jr., P.E.L. Smith, and P. Mortensen, pp. 163–188. Studies in Ancient Oriental Civilization 36, University of Chicago, Oriental Institute. Chicago.

Grigson, C. 1989. Size and Sex: Evidence for the Domestication of Cattle in the Near East. In *The Beginnings of Agriculture*, ed. A. Milles, D. Williams, and N. Gardner, pp. 77–109. BAR International

Series, 496. Oxford.
Harris, M. 1985. *Good to Eat: Riddles of Food and Culture.* Simon and Schuster, New York.
Hayden, B. 1981. Subsistence and Ecological Adaptation of Modern Hunter/Gatherers. In *Omnivorous Primates: Gathering and Hunting in Human Evolution,* ed. R.S.O. Harding and G. Teleki, pp. 344–421. Academic Press, New York.
——— 1990. Nimrods, Piscators, Pluckers and Planters: The Emergence of Food Production. *Journal of Anthropological Archaeology* 9:31–69.
——— 1992. Models of Domestication. In *Transitions to Agriculture in Prehistory,* ed. B. Gebauer and T.D. Price, pp. 11–19. Prehistory Press, Madison.
Hecker, H.M. 1975. The Faunal Analysis of the Primary Food Animals from Pre-pottery Neolithic Beidha (Jordan). Ph.D. dissertation, Columbia University, New York.
Helmer, D. 1989. Le développement de la domestication au Proche-Orient de 9500 à 7500 BP: Les nouvelles donées d'el Kowm et des Ras Shamra. *Paleorient* 15:111–145.
Hesse, B.C. 1978. Evidence for Husbandry from the Early Neolithic Site of Ganj Dareh in Western Iran. Ph.D. dissertation, Columbia University, New York.
——— 1984. These Are Our Goats: The Origins of Herding in West Central Iran. In *Animals and Archaeology.* Vol. 3, *Early Herders and Their Flocks,* ed. J. Clutton-Brock and C. Grigson, pp. 243–264. BAR International Series 202. Oxford.
Hole, F. 1984. A Reassessment of the Neolithic Revolution. *Paleorient* 10(2): 49–60.
——— 1989. A Two-Part, Two-Stage Model of Domestication. In *The Walking Larder: Patterns of Domestication, Pastoralism, and Predation,* ed. J. Clutton-Brock, pp. 97–104. Unwin Hyman, London.
Hughes, I. 1970. Pigs, Sago and Limestone: The Adaptive Use of Natural Enclosures and Planted Sago in Pig Management. *Mankind* 7:272–278.
Jarman, M.R., and P.F. Wilkinson. 1972. Criteria of Animal Domestication. In *Papers in Economic Prehistory,* ed. E.S. Higgs, pp. 83–96. Cambridge University Press, Cambridge.
Kelly, R.C. 1988. Etoro Suidology: A Reassessment of the Pig's Role in the Prehistory and Comparative Ethnology of New Guinea. In *Mountain Papuans: Historical and Comparative Perspectives from New Guinea Fringe Highland Societies,* ed. J.F. Weiner, pp. 111–186. University of Michigan Press, Ann Arbor.
Klein, R.G. 1989. Why Does Skeletal Part Representation Differ Between Smaller and Larger Bovids at Klasies River Mouth and Other Archaeological Sites. *Journal of Archaeological Science* 6:363–381.

Kuşatman, B. 1991. The Origin of Pig Domestication with Particular Reference to the Near East. Ph.D. dissertation, University College, London. University Microfilms, Ann Arbor.
Lawrence, B. 1980. Evidences of Animal Domestication at Çayönü. In *Prehistoric Research in Southeastern Anatolia,* ed. H. Çambel and R.J. Braidwood, pp. 285–308. Istanbul University Faculty of Letters Press, Istanbul.
——— 1982. Principal Food Animals at Çayönü. In *Prehistoric Village Archaeology in South-Eastern Turkey,* ed. L.S. Braidwood and R.J. Braidwood, pp. 175–199. BAR International Series 138. Oxford.
Lieberman, D.E. 1993. The Rise and Fall of Seasonal Mobility Among Hunter-Gatherers: The Case of the Southern Levant. *Current Anthropology* 34:599–631.
McArdle, J. 1990. Halafian Fauna at Girikihaciyan. In *Girikihaciyan: A Halafian Site in Southeastern Turkey,* ed. P.J. Watson and S.A. Leblanc, pp. 109–120. University of California, Los Angeles, Institute of Archaeology, Monograph 33.
Malynicz, G.L. 1970. Pig Keeping by the Subsistence Agriculturist of the New Guinea Highlands. *Search* 1:201–204.
Meadow, R.H. 1984. Animal Domestication in the Middle East: A View from the Eastern Margin. In *Animals and Archaeology.* Vol. 3, *Early Herders and Their Flocks,* ed. J. Clutton-Brock and C. Grigson, pp. 309–337. BAR International Series, 202. Oxford.
——— 1989. Osteological Evidence for the Process of Animal Domestication. In *The Walking Larder: Patterns of Domestication, Pastoralism, and Predation,* ed. J. Clutton-Brock, pp. 80–90. Unwin Hyman, London.
Meggitt, M.J. 1958. The Enga of the New Guinea Highlands: Some Preliminary Observations. *Oceania* 28:253–330.
Morren, G.E.B. 1986. *The Miyanmin: Human Ecology of a Papua New Guinea Society.* Iowa State University Press, Ames.
Payne, S. 1975. Faunal Change at Franchthi Cave from 20,000 BC to 3,000 BC. In *Archaeozoological Studies,* ed. A.T. Clason, pp. 120–131. New Holland Publishing Company, Amsterdam.
Perkins, D., Jr. 1964. Prehistoric Fauna from Shanidar. *Science* 144:1565–1566.
——— 1973. The Beginnings of Animal Domestication in the Near East. *American Journal of Archaeology* 77:279–282.
Perkins, D., Jr., and P. Daly. 1968. A Hunter's Village in Neolithic Turkey. *Scientific American* 219:97–106.
Pimental, D., and M. Pimental. 1979. *Food, Energy and Society.* Edward Arnold, London.

Redding, R.W. 1981. Decision Making in Subsistence Herding of Sheep and Goats in the Middle East. Ph.D. dissertation, University of Michigan, Ann Arbor. University Microfilms, Ann Arbor.

——— 1988. A General Explanation of Subsistence Change: From Hunting and Gathering to Food Production. *Journal of Anthropological Archaeology* 7:59–97.

——— 1991. The Role of the Pig in the Subsistence System of Ancient Egypt: A Parable on the Potential of Faunal Data. In *Animal Use and Culture Change*, ed. P.J. Crabtree and K. Ryan, pp. 20–30. MASCA Research Papers in Science and Archaeology, Supplement to Vol. 8. University of Pennsylvania Museum, Philadelphia.

——— 1994. The Vertebrate Fauna. In *Tel Anafa*. Part I:i, *Final Report on Ten Years of Excavation*, ed. S.C. Herbert, pp. 279–322. *Journal of Roman Archaeology*, Supplementary Series, no. 10.

Reed, C.A. 1960. A Review of the Archaeological Evidence on Animal Domestication in the Prehistoric Near East. In *Prehistoric Investigations in Iraqi Kurdistan*, ed. R.J. Braidwood and B. Howe, pp. 119–145. Studies in Ancient Oriental Civilization 31. University of Chicago Press. Chicago.

Rosenberg, M. 1994. Hallan Çemi Tepesi: Some Further Observations Concerning Stratigraphy and Material Culture. *Anatolica* 20:121–140.

Rosenberg, M., and M. Davis. 1992. Hallan Çemi Tepesi, an Early Aceramic Neolithic Site in Eastern Anatolia: Some Preliminary Observations Concerning Material Culture. *Anatolica* 18:1–18.

Rosenberg, M., R.M.A. Nesbitt, R.W. Redding, and T.F. Strasser. 1995. Hallan Çemi Tepesi: Some Preliminary Observations Concerning Early Neolithic Subsistence Behaviors in Eastern Anatolia. *Anatolica* 21:1–12.

Sahlins, M. 1972. *Stone Age Economics*. Aldine, New York.

Stampfli, H.R. 1983. The Fauna of Jarmo with Notes on Animal Bones from Matarrah, the Amuq, and Karim Shahir. In *Prehistoric Archaeology Along the Zagros Flanks*, ed. L.S. Braidwood, R.J. Braidwood, B. Howe, C.A. Reed, and P.J. Watson, pp. 431–483. Oriental Institute Publications, 105. Chicago.

Strathern, A. 1984. *A Line of Power*. Tavistock Publications, London.

Townsend, P.K.W. 1969. Subsistence and Social Organization in a New Guinea Society. Ph.D. dissertation, University of Michigan, Ann Arbor.

Uerpmann, H.-P. 1979. Probleme de Neolithisierung des Millelmeerraumes. Beihefte zum *Tübinger Atlas des Vorderen Orients*, Reihe B, Nr. 28. Ludwig Reichert, Weisbaden.

Williamson, G., and W.J.A. Payne. 1978. *An Introduction to Animal Husbandry in the Tropics*. Longman Group, London.

Yen, D.E. 1991. Domestication: The Lessons from New Guinea. In *Man and a Half: Essays in Pacific Anthropology and Ethnobiology in Honor of Ralph Bulmer*, ed. A. Pawley, pp. 558–569. The Polynesian Society, Auckland.

Zeder, M.A., N. Cleghorn, and H. Lapham. 1995. A Reconsideration of the Evidence for Goat Domestication in the Zagros from the Perspective of the Upper Paleolithic in Highland Iran. Paper presented at the annual meeting of the Society for American Archaeology, Minneapolis, May 6, 1995.

PIG EXPLOITATION AT NEOLITHIC ÇAYÖNÜ TEPESI (SOUTHEASTERN ANATOLIA)

Hitomi Hongo

Primate Research Institute, Kyoto University, Inuyama, Aichi 484-8506, Japan

Richard H. Meadow

Zooarchaeology Laboratory, Peabody Museum, Harvard University, 11 Divinity Avenue, Cambridge, MA 02138, USA

Introduction

Dogs, sheep, goats, cattle, and pigs are the animals that were domesticated during the Neolithic or, in the case of dogs, even earlier (e.g., Tchernov and Valla 1997; Vilá et al. 1997). During the initial phase of their domestication, sheep, goats, cattle, and pigs were kept for exploitation of primary products, that is, body parts obtained from the dead animal. Dogs generally played a different role, as camp scavengers, hunting companions, and house pets, although sometimes they were also eaten. The process of domestication of dogs was likely to have been different from that of bovids. Attracted by leftover food, wild *Canis* may have come into a commensal and then symbiotic relationship with humans. This situation provided the opportunity for close individual relationships to develop between young animals and humans (Uerpmann 1996; Tchernov and Valla 1997).

Although pigs (*Sus scrofa* L.) were (and are) exploited primarily for their meat, fat, hides, and tusks (canine teeth), relationships between wild pigs and humans at the end of the Pleistocene and beginning of the Holocene in western Asia might have had some similarities to those between wild *Canis* and humans. Boars are omnivorous and highly adaptive, as much as—or even more than—wolves (Heptner, Nasimovich, and Bannikov 1988). Scavenging on wastes from human settlements, and also attracted by cultivated plants when those began to be grown, wild pigs would have been tempted to come close to sedentary communities. In so doing, conditions for a form of mutualism would have been present, with pigs more accessible to humans both for immediate consumption and for capture. As wild pigs, like canids, are multiparous animals whose newborns are immobile for a considerable period of time, it is relatively easy to capture and raise young wild pigs.

One can argue that the exploitation of pigs can fall at one or more places along a continuum of hunting to cultural control to domestication. By hunting we mean harvesting from the wild without specific concern for individual animals. By cultural control, we mean some sort of special relationship between humans and animals that does not include breeding in captivity. Under these conditions, humans may keep individual animals, cull them selectively from free-ranging stock, or manage them in such a fashion that does not isolate breeding stock from the wild population. By domestication, we mean the keeping and breeding under conditions of captivity of individual animals and their genetic isolation from wild populations. It is important to emphasize that these human-animal relationships, sometimes dealt with as "states" or "stages," in fact represent a continuum of animal exploitation practices any number of which may (or may not) have been employed at a given time (see also Redding and Rosenberg, this volume). Given the highly adaptive nature of pig behavior, it may be difficult to identify any individual state from a record composed only of animal bone remains. This leads us to suggest that the best way to proceed is to examine a long sequence of faunal remains from a single site or region in order to document patterns of kill-off and body size variation through time. This type of investigation is possible at the Neolithic site of Çayönü in southeastern

Table 1. Çayönü periodization, dating, and quantity of faunal specimens analyzed (through 2/97). The number of fragments is not corrected for multiple specimens from a single individual. Dates provided by Mehmet and Aslı Özdoğan, University of Istanbul

Periodization of Çayönü Building subphase	Ab.	Date (B.P.)	# pieces analyzed	Analyzed Faunal Remains Wt. (g) analyzed pieces	# ID'd specimens (% ID'd)	Wt. (g) ID'd specimens (% ID'd)
Round Building	r	10,200-9,200	2,634	11,942	437 (16.6)	5,018 (42.0)
Grill Building	g	9,200-9,100	2,566	11,233	421 (16.4)	5,122 (45.6)
Channeled Building	ch	9,100-9,000	1,746	3,257	221 (12.7)	1,198 (36.8)
Cobble-paved Building	cp	9,000-8,600?	1,223	8,146	226 (18.5)	4,091 (50.2)
Cell Building	c	8,600-8,300	407	2,029	47 (11.5)	1,248 (38.5)
Large Room	lr	?8,300-8,000	0	0	0	0
Early Pottery Neolithic	P	8,000-7,500?	0	0	0	0

Anatolia, and if trends can be identified, it may be possible to suggest the existence of changing human-animal relationships along a hunting-herding continuum.

Given the great adaptability and behavioral plasticity of pigs, those measures that have been used to indicate the domestication of bovids, such as increasing proportions of remains of those taxa in the faunal record, changes in kill-off patterns, and body size diminution, may not be as clear-cut for pigs. In this paper we evaluate these traditional measures and see if any changes in human-pig interactions during the long Prepottery Neolithic sequence at Çayönü can be detected. The question of when changes in the exploitation of pigs may have taken place is of particular interest in connection with the report from the nearby site of Hallan Çemi that pigs were at least "culturally controlled" if not domesticated during the late Epipalaeolithic in a context without any evidence for plant cultivation (Redding 1994, 1995; Rosenberg 1994; Redding and Rosenberg, this volume; Rosenberg and Redding, this volume). Unfortunately, the cultural sequence at Hallan Çemi does not extend into the following Prepottery Neolithic era, so controlled comparison with later periods in the exploitation of pigs at that site cannot be made. The sequence that is lacking at Hallan Çemi is present at Çayönü (Fig. 1).

The presence of the domestic form of pigs at Çayönü was first suggested by Stampfli (in ms. from 1966; see Stampfli 1983) who observed, based on a limited review of dental remains from the 1964 season, that the ratio of domestic to wild pigs at Çayönü seemed to be about one-to-one. Berrin Kuşatman (1991) also suggested that there is some evidence for the presence of domestic pigs in the aceramic Neolithic levels of Çayönü. Until our analysis, however, pig remains from Çayönü were never studied in detail by chronological unit, and thus the timing and the nature of the appearance of domestic forms, if any, has not been evident.

The Site and the Material

Çayönü Tepesi is located about 40 km northwest of Diyarbakır in southeastern Turkey at an elevation of ca. 830 m above sea level. It is currently situated on a small tributary of the Tigris called Boğazçay. At the time of its occupation, a second stream, possibly forming a pond, may have partially encircled the site on the north. Beyond the riverine zone, the area was probably covered with an open forest consisting mainly of oak and pistachio. Such an environmental situation would have provided an ideal habitat for wild boars which, while highly adaptable, are particularly drawn to marshy locations with mixed forests that provide an abundance of food and protective cover (Heptner, Nasimovich, and Bannikov 1988).

Excavations at Çayönü were carried out between 1964 and 1991 by teams from the University of Chicago, Istanbul University, Karlsruhe University, and the University of Rome. A long sequence of occupations spanning the period between about 10,200 and 7,500 uncalibrated radiocarbon years B.P. has been uncovered (Table 1 and, e.g., Çambel and Braidwood 1980; Braidwood and Braidwood 1982; Özdoğan and Özdoğan 1990). Although

Fig. 1. Chronology of Neolithic sites in western Asia (modified from information compiled when developing Figure 3.2 in Bar-Yosef and Meadow 1995).

Pig Exploitation at Neolithic Çayönü Tepesi (Southeastern Anatolia)

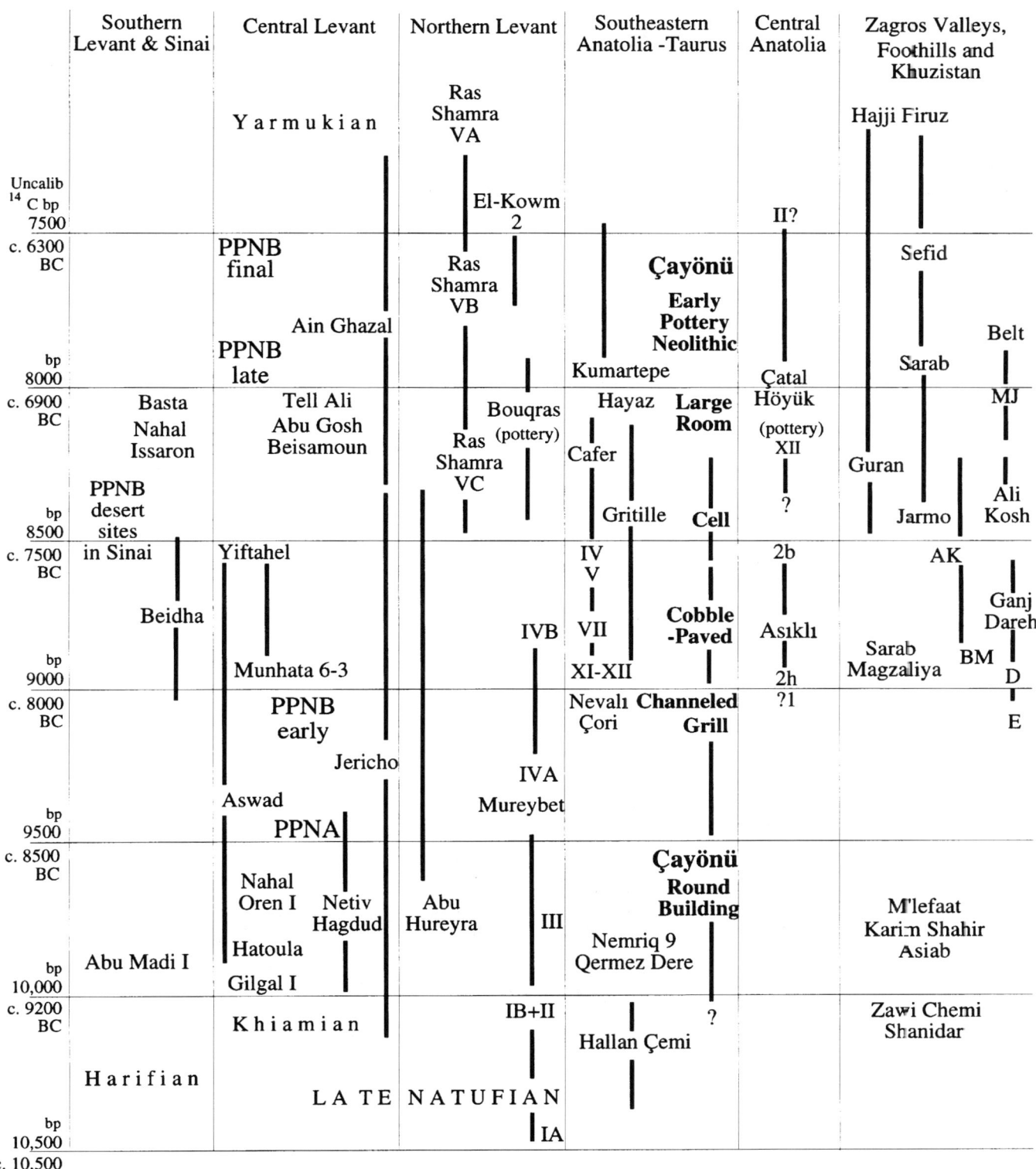

Table 2. Skeletal parts allocated to the different epiphyseal fusion stages for *Sus*. Allocations based on Silver (1969), Bökönyi (1972), Habermehl (1975), and Bull and Payne (1982). See Figure 4

Stage I	Stage II	Stage III
Distal Scapula	Proximal Phalanx 1	Proximal Humerus
Acetabulum area	Distal Metapodials	Distal Radius
Distal Humerus	Distal Tibia	Proximal and Distal Ulna
Proximal Radius	Distal Fibula	Proximal and Distal Femur
Proximal Phalanx 2	Calcaneum	Proximal Tibia

first reports on the fauna of Çayönü were published by Barbara Lawrence (1980, 1982), these deal only with a portion of the remains. Since that time, not only has new material been excavated from all periods, but understandings of the stratigraphy of the site have undergone major revisions (A. Özdoğan 1994).

Each subphase of Prepottery Neolithic Çayönü is characterized by a particular type of architecture. In chronological order, from earliest to latest, these are the Round Building, Grill Building, Channel(ed) Building, Cobble-paved Building, Cell Building, and Large Room subphases (Table 1). This Prepottery sequence is followed by the Pottery Neolithic occupation at the site. The Çayönü sequence runs parallel to the PPNA and all of the PPNB of the Levant (Fig. 1).

Re-analysis of the Çayönü faunal remains began in 1996 in Istanbul with the assistance of Gülçin İlgezdi and Banu Öksüz. Portions of the total assemblages from the Round, Grill, Channeled, Cobble-paved, and Cell Building subphases have been analyzed in detail to investigate the relative frequency of different animal taxa (Table 1). In order to investigate possible changes in pig body size and kill-off through time, additional pig remains from the Round, Grill, Channeled, and Cobble-paved subphases were recorded and measured. Finally, data collected by the late Berrin Kuşatman and included in her Ph.D. thesis (1991) have been employed when available.

Analysis

Relative proportions of taxa

Looking first at the relative representation of different taxa, Figure 2 summarizes the proportion of pig bones among the *total number of identified specimens*. *Sus* is the single most commonly encountered taxon, and its remains make up from about 30 to 40% of the identified faunal corpus during the earlier part of the Prepottery sequence in the Round, Grill, and Channeled subphases. Pigs, sheep (*Ovis* sp.), goats (*Capra* sp.), and cattle (*Bos* sp.) together make up from 60 to 70% of the faunal remains in all the subphases analyzed and from 70 to 90% of the ungulate

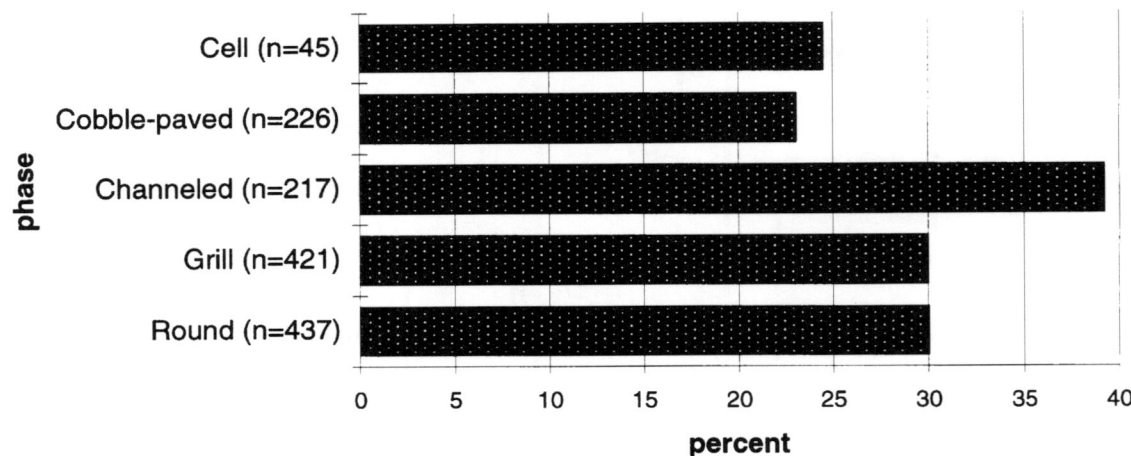

Fig. 2. Proportion of *Sus* remains among the total bones so far identified for each of the five earliest subphases at Çayönü. The number of fragments is corrected for multiple specimens from a single individual. The Round Building subphase is the oldest.

remains (Fig. 3). Additional taxa include equids (probably all *Equus hemionus*), red deer (*Cervus elaphus*), roe deer (*Capreolus capreolus*), fallow deer (*Dama dama*), gazelle (*Gazella subgutturosa*), and small mammals. Red deer is the most frequently encountered form among the cervids and particularly so in the Cobble-paved subphase, where its remains make up almost 13% of the total number of identified specimens. Among the small mammals, red fox (*Vulpes vulpes*) and hare (*Lepus capensis*) are the most commonly encountered forms.

Figure 3 summarizes the relative frequencies of bones from *animals of principal dietary importance* for humans. In this figure only the relative abundance of ungulates—pigs, bovids, cervids, and equids—are shown, with carnivores, birds, and other wild forms being excluded. Red deer, roe deer, fallow deer, gazelle, and unidentified large and medium bovids or cervids are grouped as "Cervid & Cervid/Bovid." In these reduced assemblages, pig remains make up from about 26% to nearly 60% of the quantified specimens, with the proportion being nearly 45% in the earliest two subphases and almost 60% in the next, Channeled, subphase. The frequency of pig remains decreases in the Cobble-paved and also in the Cell subphases, particularly from what it was in the Channeled subphase. (Note, however, that the sample from the Cell subphase so far analyzed is very small.) Interestingly, the Channeled subphase also has the greatest representation of sheep and goat and the least representation of cattle and cervids of all the early subphases. It represents the culmination of trends of animal exploitation begun in the Round subphase. After the Channeled subphase animal exploitation at Çayönü, as represented by the counts of identified animal bones, is transformed. This shift is characterized by a decrease in the proportion of pigs and an increase in cattle and cervids (mainly red deer).

In sum, pigs played a very important part in the diet of the Prepottery Neolithic populations of Çayönü, especially during the early part of the sequence. Indeed, their remains are much more frequent at Çayönü in the Round and Grill subphases than they are at Hallan Çemi (see Redding and Rosenberg, this volume). This lends additional impetus to an investigation of the nature of human-pig interactions during this time span.

Kill-off Patterns

Kill-off patterns for pigs in each subphase were investigated using state of epiphyseal fusion and measures of tooth eruption and wear. Post-cranial

Table 3. Tooth wear stages for *Sus*. See Grant (1975, 1982), Bull and Payne (1982), and Hongo (1996). See Figures 5 and 6

Age Stage	Tooth	Wear Stages
		(after Grant 1982)
I	dp4	a, b, c
Newborn	dp other	erupting, slight
	di	erupting, slight
II	dp4	d
M1 erupting	dp other	moderate
(up to ca. 6 months)	M1	erupting, a, b
	P1	erupting
III	dp4	e, f, g, h, i, j, k, l
M2 erupting	dp other	moderate, heavy
(ca. 6–12 months)	di	moderate, heavy
	M1	c, d, e
	M2	erupting, a, b
	M3	unerupted
	I3, C	erupting
IV	P4	a, b, c
P4 erupting	P2, P3, n.d. P	erupting, slight
(ca. 12–18 months)	I1	erupting
V	M1	f, g, h
M3 erupting	M2	c, d, e
(ca. 18–24 months)	M3	a, b
	P4	d, e
	P other	moderate
	I2	erupting
VI	M1	j, k
(over 24 months	M2	f, g, h
but not old)	M3	c, d, e
	P4	f
	P	heavy
	I	heavy
VII	M1	l, m, n
(old)	M2	j, k
	M3	f, g, h, i, j
	P4	g, h

di: deciduous incisor — I: permanent Incisor
dp: deciduous premolar — P: permanent Premolar
n.d.: unidentified — M: Molar

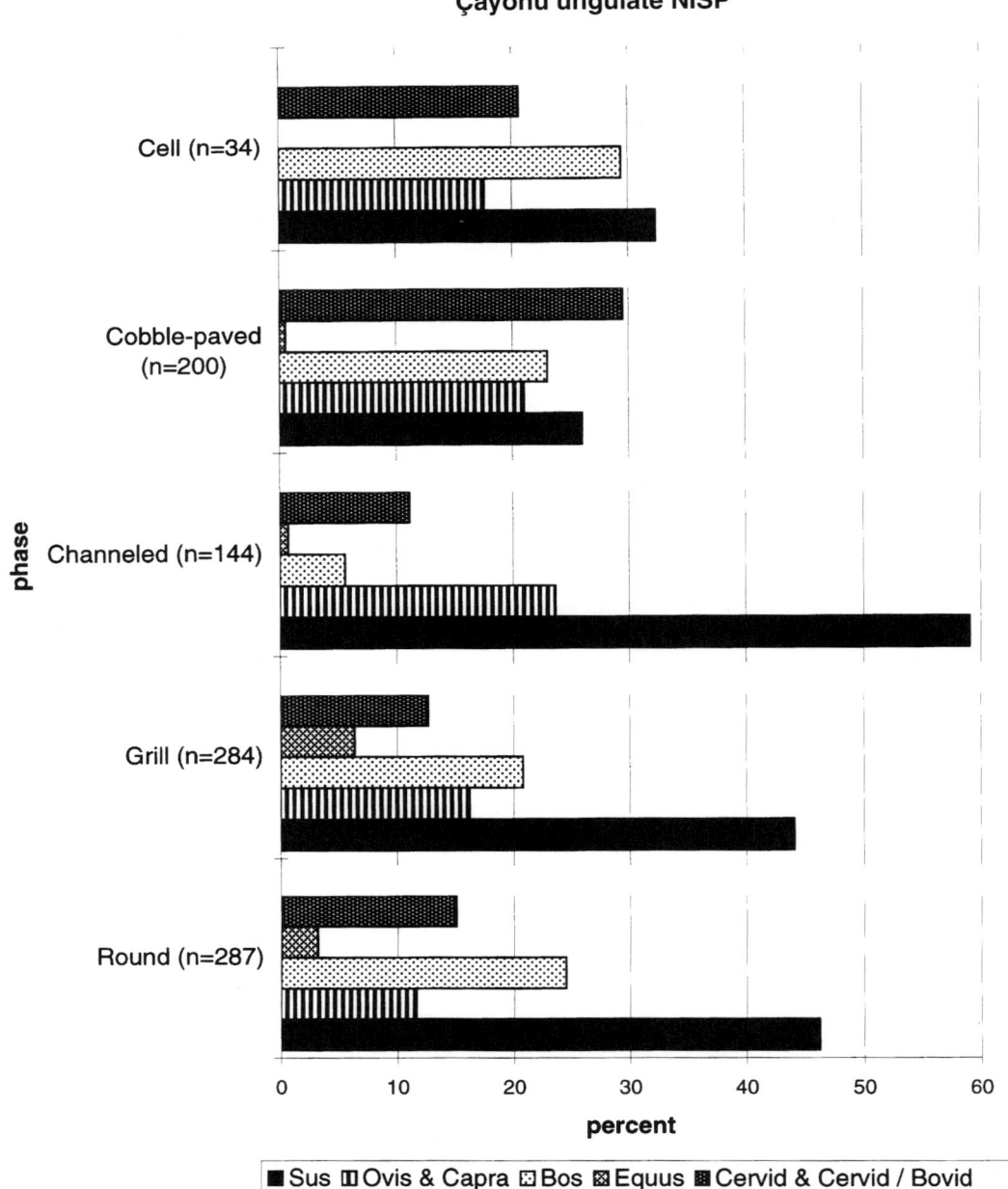

Fig. 3. Relative frequency of different ungulate taxa based on the number of identified specimens for each of the five earliest subphases at Çayönü. The Round Building subphase is the oldest. The "Cervid & Cervid/Bovid" category includes red deer (*Cervus elaphus*), roe deer (*Capreolus capreolus*), fallow deer (*Dama dama*), gazelle (probably *Gazella subgutturosa*), and unidentified large and medium bovids or cervids. The number of fragments is corrected for multiple specimens from a single individual.

parts were grouped according to the sequence of fusion presented by Silver (1969), Bökönyi (1972), Habermehl (1975), and Bull and Payne (1982) (Table 2). In general, Stage I epiphyses fuse before 12 months (infantile/juvenile), Stage II epiphyses between 24 and 30 months (subadult), and Stage III epiphyses between 36 and 42 months (adult). The age at which fusion of a particular skeletal part takes place, however, varies between populations and even within the same population. Therefore, when epiphyseal fusion data are compared between sites and even between different time periods within a site, age estimates must be applied with caution. In the discussion below, and also in the figures, only age stages are used.

Using this technique of investigation, about 55 to 70% of pigs in each of the Round, Grill, and Channeled subphases can be said to have survived Stage I, and about 45 to 55% to have survived Stage III (Fig. 4). A "rebound" during Stage II was observed for all the subphases except the Channeled. This is caused primarily by large numbers of fused distal tibiae and metapodials and may be the result of taphonomic factors or due to the selective curation or importation of some lower limb elements. Overall, kill-off patterns documented using epiphyseal fusion show few differences between the earlier three subphases, although there are progressively fewer animals appearing to survive Stage III. This trend is markedly accentuated in the Cobble-paved subphase, with only 11% of the animals appearing to survive Stage III.

In order to compare the kill-off patterns of pigs at Çayönü with those of a clearly domestic population, identically constructed "survivorship curves" for pigs at Kaman-Kalehöyük, a mound site in central Anatolia, are also plotted on Figure 4. The material from this site comes from the Middle Bronze Age (Subphase IIId-c), Late Bronze Age (IIIb-a), Early and Middle Iron Age (IId-c), and Late Iron Age (IIb-a). To judge from their osteology and morphology, almost all of the pigs exploited at Kaman were domestic (Hongo 1996). The kill-off patterns for pigs at Kaman-Kalehöyük are in stark contrast to those at Çayönü. At Kaman, only 40 to 50% of pigs survived Stage I, and fewer than 30% survived Stage III (into adulthood). Although we are well aware of the taphonomic and interpretive problems involved in using epiphyseal fusion data, survivorship of 50 to 70% of the population beyond Stage I and 45 to 55% into adulthood at Çayönü is not what we would expect for a domestic population. On the other hand, the kill-off of relatively large numbers of very young animals is *not inconsistent* with the exploitation of free-ranging pig populations. The low survivorship into adulthood in the Cobble-paved subphase, however, might indicate a shift toward an overall earlier kill-off.

Another way to investigate slaughter patterns is through analysis of tooth eruption and wear data. Loose teeth and tooth rows were classified into age stages based on wear patterns defined by Grant (1975, 1982) and Bull and Payne (1982). Stage I represents newborn animals, Stage II infantile animals (up to 6 months), Stage III juvenile animals (6 to 12 months), Stage IV and V subadult animals (12 to 18 months and 18 to 24 months, respectively), Stage VI fully adult animals (24 to 36 months), and Stage VII old animals (over 36 months) (Table 3). As is the case for epiphyseal fusion data, there are problems in relating tooth-wear stages to the calendar age at death using data obtained from modern domestic breeds. Variation in timing of eruption among modern breeds of domestic pigs is as great as six months between early maturing and late maturing breeds for some teeth (Habermehl 1975; Bull and Payne 1982: Table 1). Also, the timing of tooth eruption in wild pigs is different from that in domestic breeds. Therefore, stages and not calendar ages are used in the following discussion.

Although we are aware of the problem of small sample sizes in the Round, Grill, and Cobble-paved subphases, we can make the following observations based on tooth eruption and wear (see Fig. 5). Relatively early kill-off is indicated for the Round subphase with about half of the teeth coming from infantile or juvenile animals and with another peak at Stage V which represents animals nearing adulthood. A different pattern is observed in the Grill subphase, with only about 23% of the animals being killed during the earliest three age stages. A more reliable pattern (because of larger sample size) is that for the Channeled subphase, with about 37% of the teeth coming from young animals (Stages I to III), and another 58% from subadult animals. The pattern for the Cobble-paved subphase is somewhat similar to that for the Round subphase, with peaks at Stage III and V. It is difficult to draw any conclusions for this subphase, however, because of the small sample size. Indeed, the somewhat higher proportion of teeth in Stage IV in the Cobble-paved subphase in comparison to that in the Round subphase might indicate that the kill-off pattern in the Cobble-paved subphase is more similar to the pattern observed in the Channeled subphase, the difference being attributable to the small sample size.

The pattern observed in the Channeled subphase is the type of kill-off that might be expected in a domestic population where production for meat is being optimized, with few very young animals being slaughtered and few animals being allowed to get very old. As with epiphyseal fusion, however, kill-off patterns derived from the teeth of pigs of Kaman-Kalehöyük indicate an even earlier slaughter schedule for the domestic population there. Figure 6 (from Hongo 1996) shows that at least 50% and up to 80% of pig teeth from the Bronze and Iron Age samples come from young animals killed in Stages I, II, and III. And, again unlike Çayönü, there are some old (Stage VII) animals represented at Kaman.

When comparing the results of kill-off analysis based on epiphyseal fusion and tooth eruption and wear, it is important to understand that various factors can move the actual ages corresponding to each stage in one direction or the other. For example, epiphyseal fusion among some

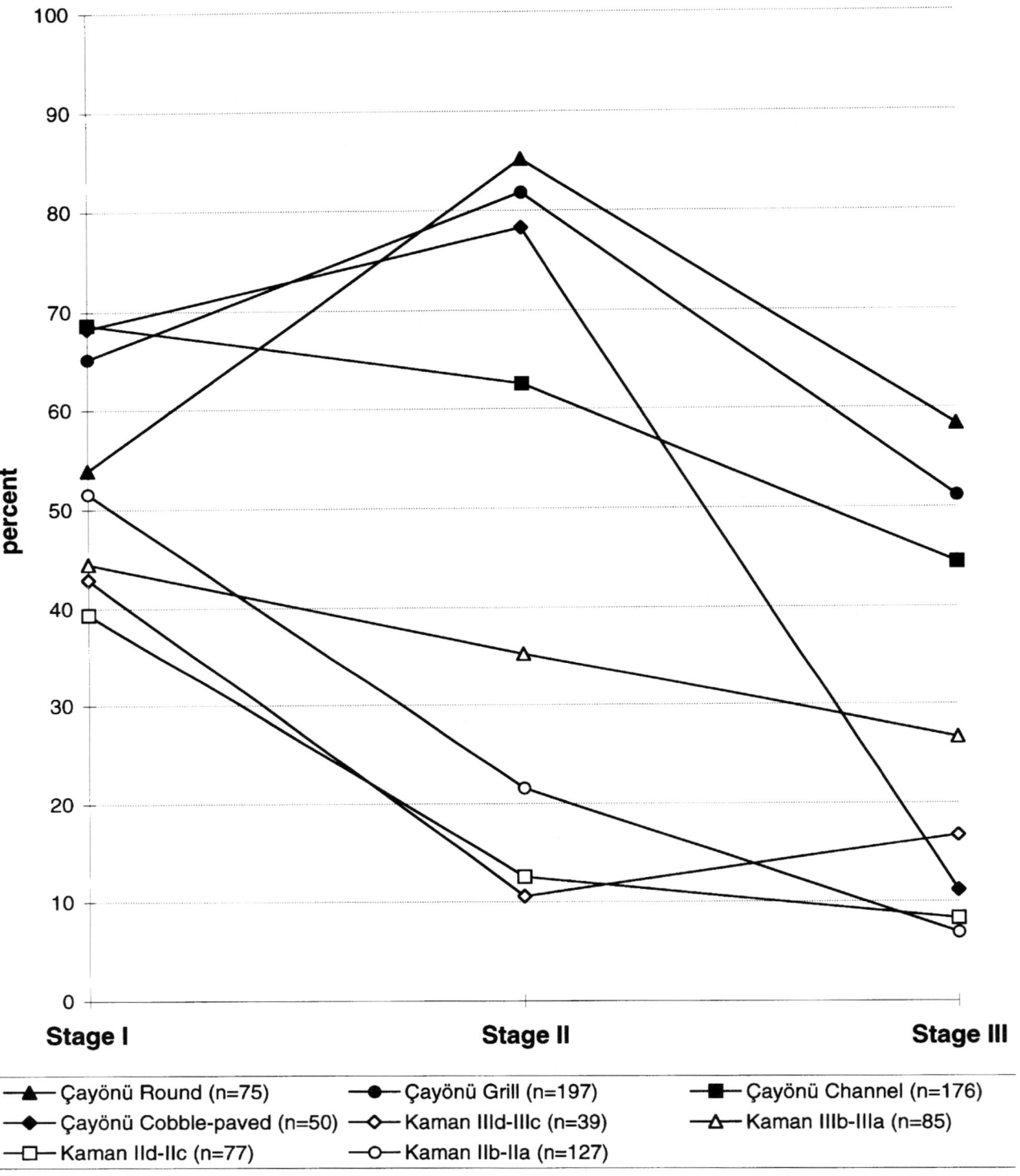

Fig. 4. Survivorship curves for *Sus* based on epiphyseal fusion for aceramic Neolithic Çayönü (Round subphase is earliest) and Bronze and Iron Age Kaman-Kalehöyük (IIId–IIIc is earliest; Hongo 1996). See Table 2 for the skeletal parts used for calculating the values for each developmental (age) stage.

ancient pig populations may have taken place at a later age than documented for modern pigs in the literature, while the attrition of teeth may have been faster due to dietary differences. Thus, the results of aging based on epiphyseal fusion and on tooth eruption and wear may not be congruent if one takes age estimates at face value. For Çayönü, overall patterns based on the two techniques are broadly coincident for the younger age stages, but are generally inconsistent for the older. Clearly, the relationships between these two methods of documenting kill-off remain to be further clarified.

Size

Size reduction in teeth is one of the characteristics used for identifying the presence of domestic pigs at a site (e.g., Flannery 1983; Stampfli 1983). Figure 7 is a plot of occlusal lengths and greatest breadths of lower third molars from various subphases of Çayönü. Many of the measurements (Table 4) were taken by Kuşatman (1991). Measurements of the lower third molars of two modern wild pigs from Turkey are also plotted for comparison. These are a female wild pig stored at the Museum of Comparative Zoology at Harvard University (specimen #51621) and a male wild pig in the first author's collection. A series of length measurements taken by Flannery (1983) on modern wild specimens from different parts of the Middle East are shown at the bottom of the chart. All of the Çayönü specimens fall in the size range for modern wild pig, although there is a trend toward somewhat smaller animals in later subphases. All the specimens from Grill and Channeled Building subphases have length measurements greater than 40 mm. Some teeth from the Cobble-paved and Cell plan subphases as well as those from the Pottery Neolithic levels fall in the range of overlap for wild and domestic pigs, between 36 and 40 mm. Since third molars come from subadult or adult animals, and these are just the animals one might expect to be better represented in a collection resulting from hunting, future analysis will need to evaluate teeth that represent younger individuals.

Turning to the post-cranial skeleton, in order to investigate the possibility of body size diminution, measurements of scapula, humerus, radius, femur, tibia, astragalus, calcaneum, and metapodials were compared to the corresponding dimensions of a "standard animal" using the "difference of logs" technique developed by Meadow (1981). From the base-10 logarithm of the measurement of an archaeological specimen is subtracted the base-10 logarithm of the corresponding dimension of the standard animal. The resulting difference is plotted on a graph. This procedure is the same as dividing each measurement of an archaeological specimen by the corresponding dimension of the standard animal and permits dimensions of different parts of the body to be plotted on the same graph. The underlying assumption is that allometric relations in the animals that yielded the archaeological specimens were similar to those in the standard animal. To make this more likely, measurements of the female Turkish wild pig skeleton stored at the Museum of Comparative Zoology, Harvard University (specimen #51621), were used as the standard (Table 5). This animal is near the small end of the size range of modern Turkish wild pigs. Even so, in the future when our analysis is complete, element by element examination of pig bone dimensions will be undertaken to test the results of the log difference approach.

In Figure 8a–d, length measurements and breadth or depth measurements are dealt with separately and not combined into one set of histograms. This is because the length of a bone is related to the height of the animal and not so directly to the weight (mass) of the animal, which is reflected in the breadth and depth dimensions. If more than one length or breadth/depth measurement was obtained from a specimen, the average value for the log size indices was used. When both lengths and breadths/depths were available, the log size indices for breadth or depth measurements were plotted in scatter-plots against those for the length measurements of the same specimen, so that changes in proportions of individual animals, if any, can be observed. In many cases, however, it was not possible to take length measurements and in some cases, not the breadth or depth measurements. Those single dimensions were combined with the others and displayed as histograms (with intervals of 0.020) to the left of (lengths) and below (breadths or depths) the scatter plots.

Most of the pig bones in the Round plan subphase (Fig. 8a) are larger than the corresponding elements of the standard animal, although a few smaller specimens also exist. In the Grill subphase, more bones from small animals are present (Fig. 8b), resulting in more variability in size, although these are breadth/depth measurements without corresponding lengths. In the Channeled and Cobble-paved Building subphases, the tendency continues towards there being a few smaller (especially lighter weight) animals (Fig. 8c, d), although the number of specimens for the Cobble-paved subphase is very small.

In order to be able to compare the size of pigs from Çayönü with later domestic pigs in Anatolia, log size indices for pig bones from Kaman-Kalehöyük in the second and first millennium B.C. are plotted in Figure 9 using the same standard animal (MCZ #51621). Because we do not have good information on size variability in wild Turkish pigs nor on the degree of sexual dimorphism in domestic pigs in Anatolia, we can say only that one measured specimen from Kaman-Kalehöyük is, with a high degree of probability, from a wild pig. And if we tentatively define the lower limit of a likely area of overlap between wild and domestic pigs to be about –0.075, there are some breadth measurements from Kaman that fall into this range of overlap and therefore could be wild females. Using this same measure, the smallest specimens in the Round subphase at Çayönü are almost all larger than the domestic pigs from the second and first millennia B.C. As noted, however, starting from the Grill subphase some smaller specimens are observed, especially in breadth/depth dimensions.

Figure 10 summarizes the log size index data for pigs from Çayönü and Kaman-Kalehöyük. The "battleship curves" show the percentage of indices falling into the various 0.020 intervals for lengths and breadth/depths from each temporal division, with the medians also being plotted. There is considerably more overall variability in the breadth/depth dimensions than there is in the length dimensions. Nevertheless, a small overall shift toward less heavy animals from the Round through the Grill to the Channeled subphase is indicated by smaller median values, which reflect the overall lower center of gravity of the battleship curves. While there are also a few quite small specimens in the Grill and Channeled subphases, in no case does the median fall below 0.00. Indeed, there is even something of a rebound in the Cobble-paved subphase toward overall larger animals—although the smallest specimens (in breadth/depth) also occur in this subphase. To be particularly noted is the marked differences between the Çayönü and Kaman-Kalehöyük assemblages. This contrast shows the degree to which pigs were to change in size over the next millennia. Future work on the Çayönü material will include documentation of collections from the Cell, Large Room, and Pottery Neolithic to see if there is any evidence for size diminution continuing in those subphases at the site.

Looking at these data in another way, if we postulate an area of overlap between wild boar and domestic pigs to be from log index 0.00 to log index –0.075, then for the Grill, Channeled, and Cobble-paved subphases at Çayönü there are only 8 dimensions out of 323 (2.5% of the plotted points) that fall below this overlap area. Only 4 of these are considerably smaller (less than log index –0.10): 3 in the Grill and 1 in the Cobble-paved subphase. If the area of overlap is decreased to between 0.00 and –0.05, the number of smaller dimensions climbs to only 13 or 4.0%. Thus while there are certainly a few small animals represented, the picture is not a clear-cut one, particularly given our lack of knowledge about variability in Anatolian wild boar populations both now and in the past. On the one side there could have been "natural" size diminution in boars over the course of the Holocene in accord with general patterns of post-Pleistocene size diminution, due at least in part to pressure on the best habitats by expanding human communities. On the other side there could have been size diminution due to the escape, cross-breeding, and/or feralization of domestic stock. Also there may have been considerably greater variability in the natural wild boar populations of the time than we have any good idea of. As it is, we know that there were some extremely large males in the region. These are attested by the large M/3 from the Grill subphase (Fig. 7) and by an even larg-

Table 4. Occlusal lengths and greatest breadths of *Sus* mandibular third molars from Çayönü, and from a modern wild Turkish male (Hitomi Hongo collection) and female (MCZ #51621)

Subphase	Length	Breadth
Grill Building	48.1	21.5
Grill Building	44.0	20.2
Channeled Building	40.4	17.0
Channeled Building	41.0	17.3
Channeled Building	44.5	17.8
Cobble-paved Building*	45.2	19.7
Cobble-paved Building*	40.0	19.2
Cobble-paved Building*	38.2	19.9
Cobble-paved Building*	40.7	17.3
Cobble-paved Building*	36.0	18.7
Cobble-paved Building*	44.5	20.2
Cobble-paved Building*	43.0	19.3
Cell Building	40.0	18.7
Cell Building*	40.4	20.0
Cell Building*	37.8	17.8
Cell Building*	39.8	17.7
Cell Building*	41.0	18.0
Cell Building*	43.6	19.5
Cell Building*	41.2	21.8
Cell Building*	41.0	20.0
Cell Building*	42.5	18.7
Pottery Neolithic*	38.5	18.0
Pottery Neolithic*	39.4	18.5
Pottery Neolithic*	38.0	18.8
Modern Turkish female (MCZ #51621)	38.5	18.5
Modern Turkish male (H.H. collection)	39.2	18.8

*after Kuşatman (1991)

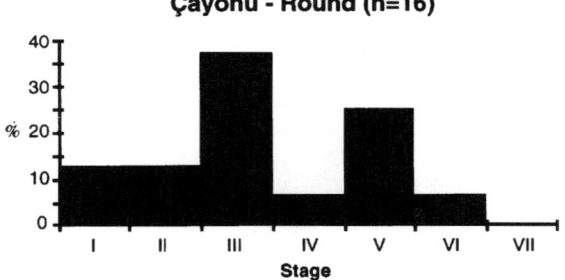

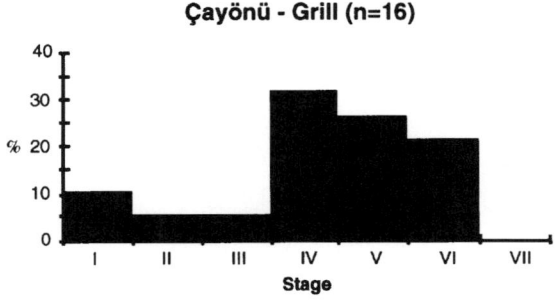

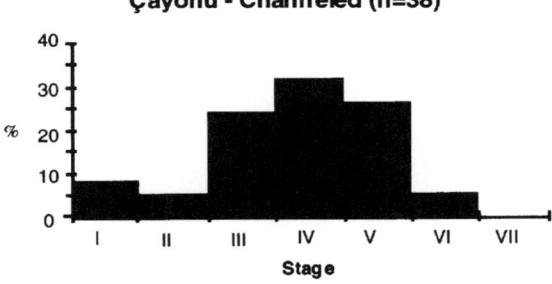

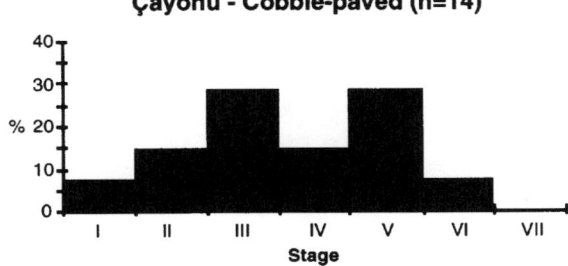

Fig. 5. Kill-off patterns for *Sus* from aceramic Neolithic Çayönü based on tooth eruption and wear. See Table 3 for definitions of tooth eruption and wear stages (which are *not* equivalent to epiphyseal fusion stages). The Round Building subphase is the oldest.

er specimen from Pottery Neolithic Hajji Firuz in northwestern Iran (Meadow 1983). At the other end of the size continuum, we just do not know how small wild females could have been.

Discussion and Conclusion

High proportions of pig bones in the earlier Prepottery Neolithic levels at Çayönü suggest that pigs were exploited intensively. These pigs may have been attracted by crops around the site and by the chance to scavenge on wastes from the village. A mutualistic relationship between humans and pigs may have already existed in the Round Building subphase. However, based on the samples analyzed to date, we cannot characterize the Çayönü pig material in any simple fashion. The kill-off patterns and the size of pigs at Çayönü as reflected in the faunal remains do not show the unequivocal characteristics of a fully domestic population. Both epiphyseal fusion and tooth eruption and wear data indicate that about 50% or more of the pigs at Çayönü survived into the subadult or adult stages. None of the third molars from Çayönü are smaller than the range of overlap of wild and domestic pigs. The vast majority of log size indices are larger than those for a relatively small modern wild female, although a few smaller animals are also represented. When the body size and kill-off patterns of pigs from Çayönü are compared with those of the fully domestic population at Bronze and Iron Age Kaman-Kalehöyük, the differences are striking (e.g., Figs. 4 and 10).

There are, however, general trends in the analyzed pig data toward features that can be considered characteristic of domestic populations. Thus the presence of a few specimens from small animals and a slightly earlier kill-off in later subphases might be significant. If we had to identify a point when we might be able to make a case for at least some pigs being kept in the community, it would be perhaps as early as the Grill subphase, but certainly by the Channeled subphase and continuing into the Cobble-paved subphase. We say "kept" here advisedly, because size reduction in individual animals can be caused by malnourishment from an early age. It is possible, therefore, that animals captured when young and kept alive for some time in the village could be represented by smaller skeletal elements in the archaeological record.

From the data at hand, it appears likely that free-ranging populations of pigs continued to be exploited at Çayönü through at least the Cobble-paved phase. How such exploitation was carried out and whether there was a breeding stock genetically isolated from the wild population remain unclear. Perhaps increasing sample sizes for all subphases will permit the identification of more distinct trends. It is also necessary to investigate whether

Table 5. Standard measurements for *Sus*. These are averaged left and right measurements (in millimeters) of a female wild boar killed near Elazığ in southeastern Turkey. The skeleton is specimen number 51621 in the collection of the Mammal Department, Museum of Comparative Zoology, Harvard University. Measurements were taken by the second author, assisted by Ajita K. Patel

Bone													
Scapula	Definition	SLC	GLP	LG	BG	HS	DHA	Ld	SBC				
	Measurement	26.5	39.4	32.5	27.5	228.1	229.6	127.9	12.7				
Humerus	Definition	Bd	BT	BFT	SD	Bp	Dp	GLC	GL	Dd	GLT	LT(mid)	LT(lat)
	Measurement	45.9	37.5	34.6	17.8	58.5	74.9	207.2	232.3	46.2	32.4	21.3	23.7
Radius	Definition	BFp	DFp	SD	Bd	BFd	DFd	GL					
	Measurement	34.2	22.7	19.0	39.2	33.0	21.0	177.0					
Ulna	Definition	BPC	DPA	SBO	SDO	GL	LPA	LO	SLFp				
	Measurement	25.3	42.4	11.5	32.2	240.9	29.9	72.4	21.5				
Metacarpal II	Definition			BFd	DFd	GL							
	Measurement			12.5	16.5	64.5							
Metacarpal III	Definition	Bp	SD	BFd	DFd	GL							
	Measurement	20.7	14.6	19.1	19.7	86.8							
Metacarpal IV	Definition	Bp	SD	BFd	DFd	GL							
	Measurement	19.1	14.6	17.7	19.4	87.8							
Metacarpal V	Definition			BFd	DFd	GL							
	Measurement			13.7	17.6	63.8							
Femur	Definition	GL	GLC	SD	Bp	Dp	DC	Bd	B.tr.pat.	Dd			
	Measurement	251.4	250.2	21.0	66.8	37.8	29.8	53.2	26.7	64.4			
Tibia	Definition	GL	SD	Bp	Dp	Bd	Dd	BFd					
	Measurement	231.3	20.6	56.7	57.4	33.5	30.6	25.4					
Astragalus	Definition	GLl	GLm	LA	Dl	Bd	Bp						
	Measurement	47.5	43.6	38.7	25.0	27.6	24.0						
Calcaneum	Definition	GL	GB	GDl	SDTc	LTc	Ld	LF Om.					
	Measurement	95.4	26.8	34.1	22.0	63.5	37.6	13.1					
Metatarsal II	Definition			BFd	DFd	GL							
	Measurement			11.4	16.2	69.0							
Metatarsal III	Definition	Bp	SD	BFd	DFd	GL							
	Measurement	17.9	13.3	18.4	20.3	97.2							
Metatarsal IV	Definition	Bp	SD	BFd	DFd	GL							
	Measurement	17.5	14.3	18.5	21.3	105.4							
Metatarsal V	Definition			BFd	DFd	GL							
	Measurement			11.7	17.0	73.6							

Abbreviations as in Driesch (1976) with additions as noted below.
Definitions follow Kuşatman (1992) based on Driesch (1976) with the following additions/clarifications:
  Humerus BFT: as with BT but includes only articular surface; Humerus GLT, LT(mid), LT (lat): greatest length of trochlea (medially) & least length in the middle & laterally;
  Ulna SLFp (smallest length of faces articularis proximalis): least length (diameter) of articular surface for humerus;
  Femur B.tr.pat. (Breadth of the trochlear patellaris): breadth of the articular surface for the patella;
  Astragalus LA: from the groove between the condyles of the proximal end to the most distal point on the distal end;
  Calcaneum SDTc & LTc: least depth of Tuber calcaneum and length of Tuber calcaneum on dorsal side;
  Calcaneum Ld: length of the distal portion from process for Os malleolare to the distal end;
  Calcaneum LF Om.: length of articular process for Os malleolare.

the trends that have been observed in the early subphases continue through the following Cell, Large Room, and Pottery Neolithic temporal divisions.

This study suggests that measures to investigate domestication of bovids may not be useful in pigs, especially if genetic isolation from the wild population was not as strictly established for pigs as for bovids at an early stage of their being kept by humans. Perhaps the keeping of pigs may be better identified by changes in details of skull and mandibular morphology like those that are used to indicate the keeping of dogs. It is also possible, however, that investigation of morphological and even demographic indicators may not permit us to identify some kinds of relationships between humans and some of the animals with which they maintained close ties.

Some have argued that pigs may initially have been kept as they still are in parts of island Southeast and East Asia and Oceania (e.g., Rappaport 1967 for one people in New Guinea; see also Redding and Rosenberg, this volume). Females would be penned at night and allowed to roam freely during the day, attracted back to the settlement by food provided by their owners. Breeding would take place with feral males, thus not creating an isolated gene pool. S. Hayashida (1971) has suggested, based on his research on indigenous pig breeds in Southeast Asia and southern Japan, that such a practice might be the original form of pig keeping. He reports that in a small village on Iriomote Island in southern Japan, only a few domestic pigs are kept, because wild pigs are abundant in the area and hunters can kill about 500 wild pigs a year, which is enough to support all 20 households of the village. When the villagers' domestic female pigs are in heat, male wild pigs roam around the village, and sometimes the female pigs breed with the wild pigs. It is also a common practice there to capture and raise young wild pigs.

Another possibility is that pigs might not be penned at all as in some towns in India where they subsist off human refuse throughout the settlement and are collected for slaughter on a periodic basis by those with rights to do so (RHM personal observations). In both cases, pigs are very much part of the settled landscape and they are relatively small in size. Yet pigs and especially wild boars can be dangerous animals, and they can be particularly destructive to ripening crops. In Southeast Asia horticultural plots are fenced against pigs, while in India the animals are walled out or kept away from agricultural areas. In Japan, agricultural fields in mountainous areas are often surrounded by fences and pit-traps, with the wild pigs so captured being killed and eaten. Another way of keeping pigs is briefly described in a history book written in eighth century A.D. Japan. It states that in the early fifth century pigs were brought into an open field called *Ikaino* (pig-keeping field) and were set loose. A specialized pig-keeping caste managed them (Ikata 1940; Kamo 1976), and there was clear ownership of the stock. It is not clear in this document whether the pigs described were captured wild pigs or domestic stock, although a statement in this text that the pigs were brought long distances from western Japan makes the latter seem more likely. There may or may not have been a fence surrounding the field. In this case, pigs were kept away from settled villages, and interactions between pigs and humans were probably minimal except when pigs were rounded up and culled on a periodic basis.

Whether such models of pig raising are useful for understanding the ancient practice in southeastern Anatolia can be debated. There are really no archaeological data available at the present time that are suitable to test their implications. What we can say from the evidence we do have from Çayönü is that pigs were not exploited in the way that they were later and that size diminution had not progressed to any great extent. We believe that pigs were hunted and some animals perhaps brought under some form of "cultural control" in the sense that individual animals were kept or that the exploited pig population was managed in such a way that the breeding stock was not isolated from the wild population. And here we are at the crux of the matter: individual pigs versus pig populations. Zooarchaeological techniques of investigating kill-off present overall patterns of exploitation of a "population," which is defined as those animals that are represented by their bones in an archaeological record that is not complete and that covers a significant length of time. This is true also of the way that measurement data can be employed. But measurement data can also be used to evaluate the condition of individual animals, as can morphological data. In this paper we have been dealing with the pig "populations" of Prepottery Neolithic Çayönü. In the future it will be necessary to look also at what the bones of individual animals might be able to tell us.

Acknowledgments and Dedication

Research on the Çayönü faunal material is being carried out under grants from the Nissan Science Foundation (Japan) and from the National Science Foundation (USA). We wish to thank the following individuals for assistance and support (in alphabetical order): Linda Braidwood, Robert S. Braidwood, Halet Çambel, Barbara Lawrence, Aslı Özdoğan, and Mehmet Özdoğan. We particularly wish to acknowledge the analytical help provided by Gülçin Igezdi and Banu Öksüz (University of Istanbul), and Professor Ufuk Esin for supporting their participation.

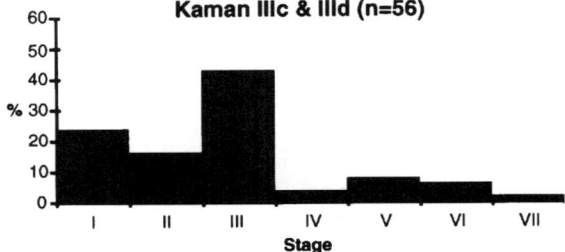
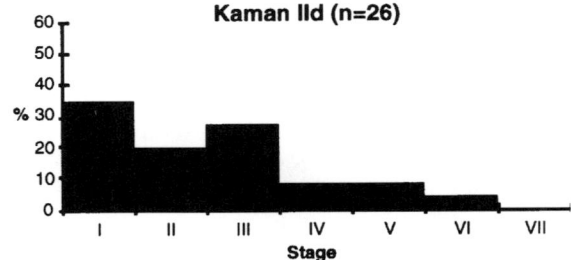
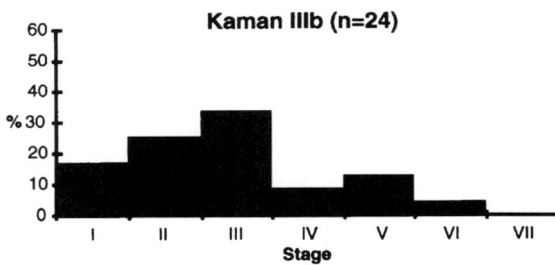
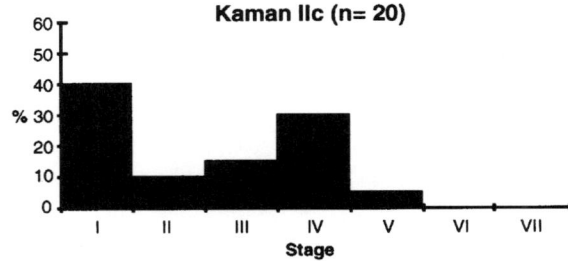
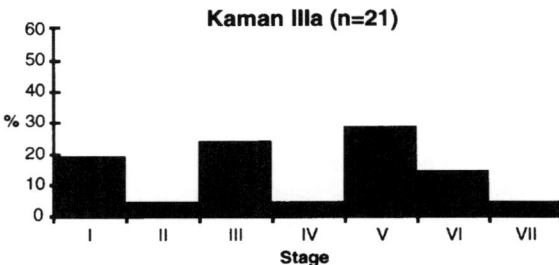
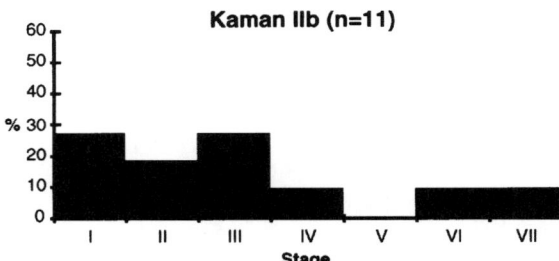

This paper is dedicated to the memory of Barbara Lawrence and Berrin Kuşatman whose spirits continue to accompany us as we investigate animal exploitation patterns at Neolithic Çayönü.

References

Bar-Yosef, O., and R.H. Meadow. 1995. The Origins of Agriculture in the Near East. In *Last Hunters, First Farmers*, ed. T.D. Price and A.B. Gebauer, pp. 39–94. School of American Research Press, Santa Fe.

Bökönyi, S. 1972. Zoological Evidence for Seasonal or Permanent Occupation of Prehistoric Settlements. In *Man, Settlement and Urbanism*, ed. P.J. Ucko, R. Tringham, and G.W. Dimbleby, pp. 121–126. Duckworth, London.

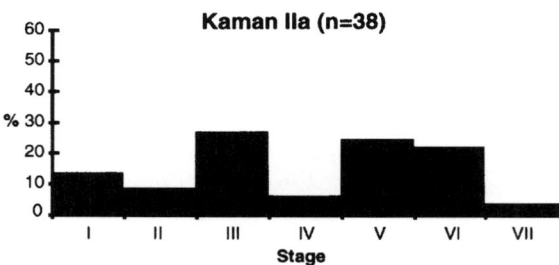

Fig. 6. Kill-off patterns for *Sus* from Bronze and Iron Age Kaman-Kalehöyük based on tooth eruption and wear (from Hongo 1996). See Table 3 for definitions of tooth eruption and wear stages (which are *not* equivalent to epiphyseal fusion stages). Substages IIIc and IIId are the oldest, IIa the youngest.

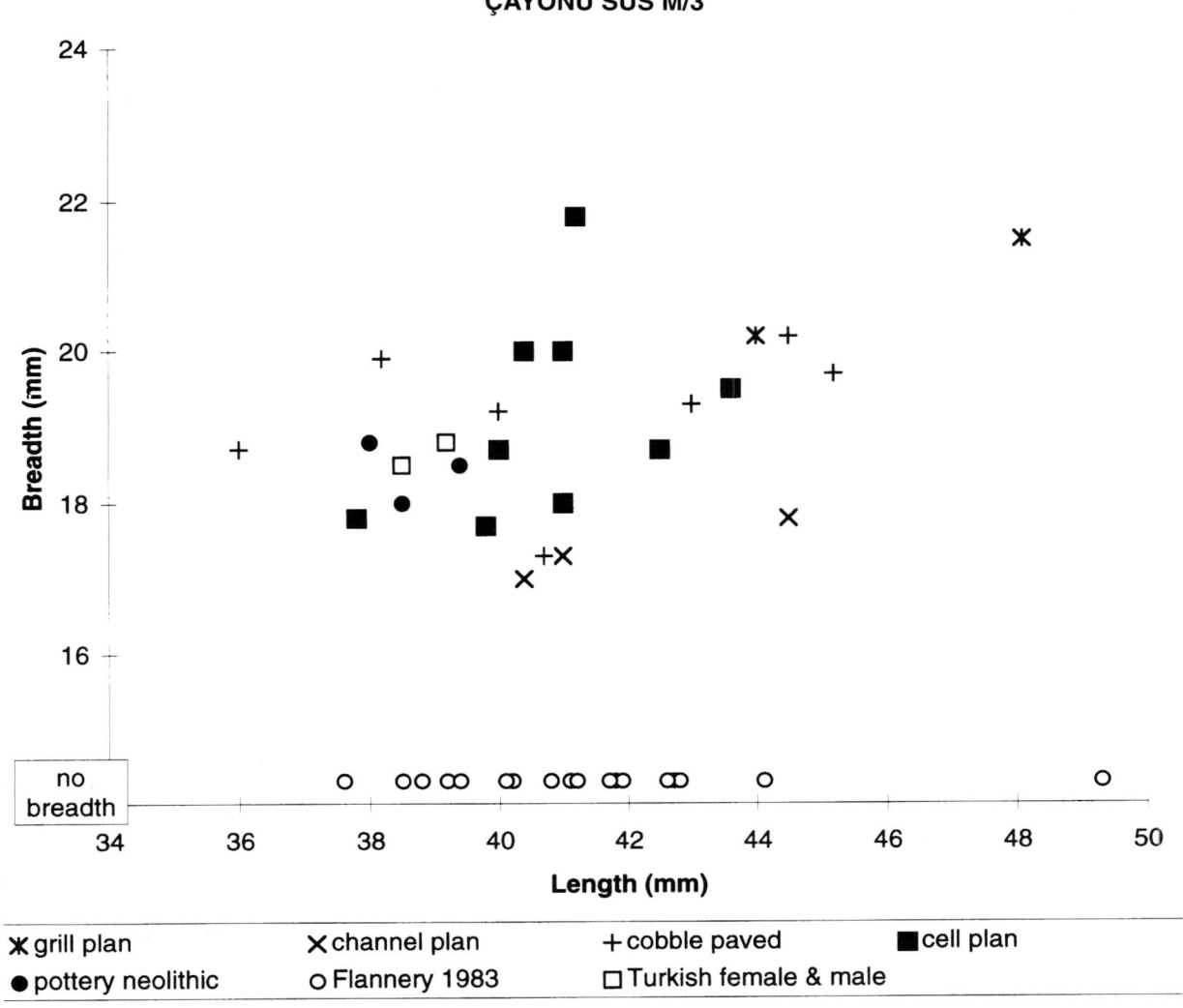

Fig. 7. Occlusal lengths and greatest breadths (where available) of *Sus* mandibular third molars from Çayörü, from a modern wild Turkish male (Hongo collection) and female (MCZ #51621), and from modern specimens (lengths only) reported by Flannery (1983). The area of overlap between wild and domestic pig is considered to be between 36 and 40 mm (Flannery 1983; Stampfli 1983).

Braidwood, L.S, and R.J. Braidwood (eds.). 1982. *Prehistoric Village Archaeology in South-Eastern Turkey.* BAR International Series 138. Oxford.

Bull, G., and S. Payne. 1982. Tooth Eruption and Epiphysial Fusion in Pigs and Wild Boar. In *Ageing and Sexing Animal Bones from Archaeological Sites*, ed. B. Wilson, C. Grigson, and S. Payne, pp. 55–71. BAR British Series 109. Oxford.

Çambel, H., and R.J. Braidwood (eds.). 1980. *İstanbul ve Chicago Üniversiteleri Karma Projesi Güneydoğu Anadolu Tarihöncesi Araştır maları I / The Joint Istanbul-Chicago Universities Prehistoric Research in Southeastern Anatolia I.* İstanbul Üniversitesi Edebiyat Fakültesi Yayınları no. 2589. Istanbul.

Driesch, A. von den. 1976. *A Guide to the Measurement of Animal Bones from Archaeological Sites.* Peabody Museum Bulletin 1. Peabody Museum, Harvard University, Cambridge.

Flannery, K.V. 1983. Early Pig Domestication in the Fertile Crescent: A Retrospective Look. In *The Hilly Flanks and Beyond: Essays on the Prehistory of Southwestern Asia*, ed. T.C. Young, Jr., F.E.L. Smith, and P. Mortensen, pp. 163–188. Studies in Ancient Oriental Civilization, vol. 36. The Oriental Institute of the University of Chicago, Chicago.

Grant, A. 1975. The Animal Bones. In *Excavations at Portchester Castle. I: Roman*, ed. B.W. Cunliffe, pp. 378–408. Society of Antiquaries, London.

——— 1982. The Use of Tooth Wear as a Guide to the Age of Domestic Ungulates. In *Ageing and Sexing Animal Bones from Archaeological Sites*, ed. B. Wilson, C. Grigson, and S. Payne, pp. 91–108. BAR British Series 109. Oxford.

Habermehl, K.-H. 1975. *Die Altersbestimmung bei Haus- und Labortieren*. 2 Auflage. Verlag Paul Parey, Berlin.

Hayashida, S. 1971. Inoshishi to Buta Soshite Nihon Minzoku (Wild Pigs, Domestic Pigs, and the Japanese). *Kokogaku Journal* 52:11–15. In Japanese.

Heptner, V.G., A.A. Nasimovich, and A.G. Bannikov. 1988. *Mammals of the Soviet Union: Volume 1, Artiodactyla and Perissodactyla*. Amerind Publishing Co., New Delhi.

Hongo, H. 1996. *Patterns of Animal Husbandry in Central Anatolia from the Second Millennium BC through the Middle Ages: Faunal Remains from Kaman-Kalehöyük, Turkey*. Ph.D. dissertation, Dept. of Anthropology, Harvard University, Cambridge.

Ikata, S. 1940. *Nihon Kodai Kachiku-shi (History of Domestic Animals in Ancient Japan)*. Kawade-shobo, Tokyo. In Japanese.

Kamo, G. 1976. *Nihon Chikusan-shi (History of Stock Raising in Japan)*. Hosei University Press, Tokyo. In Japanese.

Kuşatman, B. 1991. *The Origins of Pig Domestication with Particular Reference to the Near East*. Ph.D. dissertation, Institute of Archaeology, University College London.

Lawrence, B. 1980. Evidences of Animal Domestication at Çayönü. In *İstanbul ve Chicago Üniversiteleri Karma Projesi Güneydoğu Anadolu Tarihöncesi Araştır maları I / The Joint Istanbul-Chicago Universities Prehistoric Research in Southeastern Anatolia I*, ed. H. Çambel and R.J. Braidwood, pp. 285–308. İstanbul Üniversitesi Edebiyat Fakültesi Yayınları no. 2589. Istanbul.

——— 1982. Principal Food Animals at Çayönü. In *Prehistoric Village Archaeology in South-Eastern Turkey*, ed. L.S. Braidwood and R.J. Braidwood, pp. 175–199. BAR International Series 138. Oxford.

Meadow, R.H. 1981. Early Animal Domestication in South Asia: A First Report of the Faunal Remains from Mehrgarh, Pakistan. In *South Asian Archaeology 1979*, ed. H. Härtel, pp. 143–179. Dietrich Reimer Verlag, Berlin.

——— 1983. The Vertebrate Faunal Remains from Hasanlu Period X at Hajji Firuz. Appendix G in *Hajji Firuz Tepe, Iran: The Neolithic Settlement* by M.M. Voigt. Hasanlu Excavation Reports, Vol. I, pp. 369–422. The University of Pennsylvania Museum, Philadelphia.

Özdoğan, A. 1994. *Çayönü Yerleşmesinin Çanak Çömleksiz Neolitikteki Yeri*. Ph.D. dissertation, Faculty of Letters, Istanbul University, Istanbul.

Özdoğan, M., and A. Özdoğan. 1990. Çayönü: A Conspectus of Recent Work. In *Préhistoire de Levant II*, ed. O. Aurenche and M.C. Cauvin, pp. 387–396. Editions du CNRS, Lyon.

Rappaport, R.A. 1967. *Pigs for the Ancestors: Ritual in the Ecology of a New Guinea People*. Yale University Press, New Haven.

Redding, R.W. 1994. The Evolution of Human Subsistence Behavior and Food Production in Southern Anatolia. Paper presented at the Society for American Archaeology meeting in Anaheim, California.

——— 1995. A Piece of the Flock: Early Domestic Animals as Insurance. Paper presented at the Society for American Archaeology meeting in Minneapolis, Minnesota, May 3–7, 1995.

Rosenberg, M. 1994. Hallan Çemi Tepesi: Some Further Observations Concerning Stratigraphy and Material Culture. *Anatolica* 20:121–140.

Silver, I.A. 1969. The Ageing of Domestic Animals. In *Science in Archaeology*, 2d ed., ed. D. Brothwell and E.S. Higgs, pp. 283–302. Thames and Hudson, London.

Stampfli, H.R. 1983. The Fauna of Jarmo with Notes on Animal Bones from Matarrah, the 'Amuq, and Karim Shahir. In *Prehistoric Archaeology Along the Zagros Flanks*, ed. L.S. Braidwood, R.J. Braidwood, B. Howe, C.A. Reed, and P.J. Watson, pp. 431–483. Oriental Institute Publications 105. The Oriental Institute of the University of Chicago, Chicago.

Tchernov, E., and F.F. Valla. 1997. Two New Dogs, and Other Natufian Dogs, from the Southern Levant. *Journal of Archaeological Science* 24:65–95.

Uerpmann, H.-P. 1996. Animal Domestication—Accident or Intention? In *The Origins and Spread of Agriculture and Pastoralism in Eurasia*, ed. D.R. Harris, pp. 227–237. Smithsonian Institution Press, Washington, DC.

Vilá, C., P. Savolainen, J.E. Maldonado, I.R. Amorim, J.E. Rice, R.L. Honeycutt, K.A. Crandall, J. Lundeberg, and R.K. Wayne. 1997. Multiple and Ancient Origins of the Domestic Dog. *Science* 276:1687–1689.

Fig. 8a-d. Çayönü *Sus* log size index diagrams based on the technique outlined by Meadow (1983). The "0"-line or "standard animal" is a modern female Turkish wild boar (MCZ #51621), measurements for which are listed in Table 4. Points to the left of the y-axis represent specimens that are less broad (medio-laterally) or less deep ("anterio-posteriorly") than the standard; points below the x-axis represent specimens that are shorter (proximo-distally) than the standard. The arrows along each axis denote the median points. Histograms along the bottom and left sides of the scatter-plots display length (to the left) and breadth/depth (to the bottom) data for specimens plotted in the scatter-plots as well as for additional specimens that could not be measured in both planes. The histogram interval is 0.020.

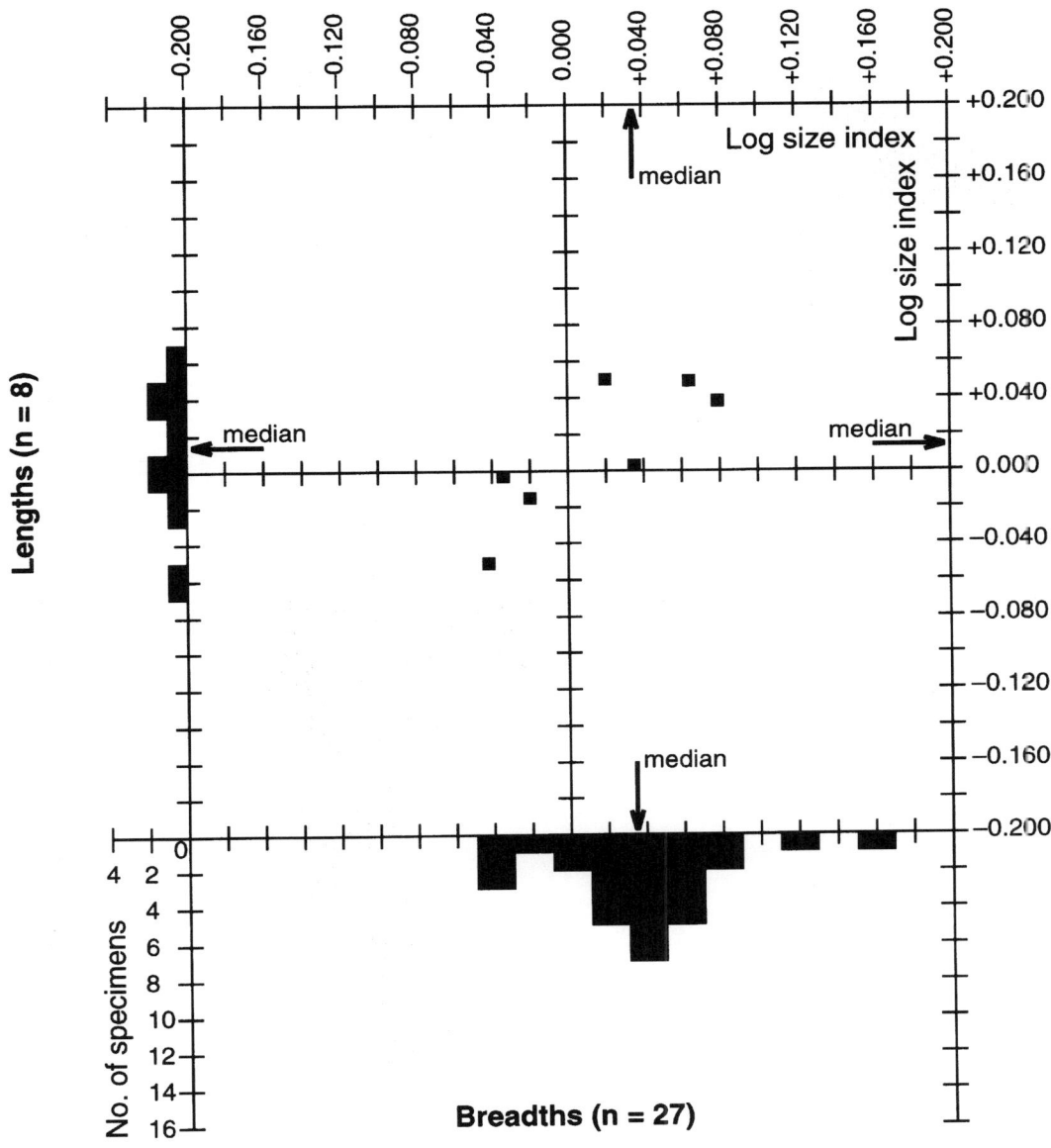

Fig. 8a. Çayönü *Sus*. Round Building subphase (earliest).

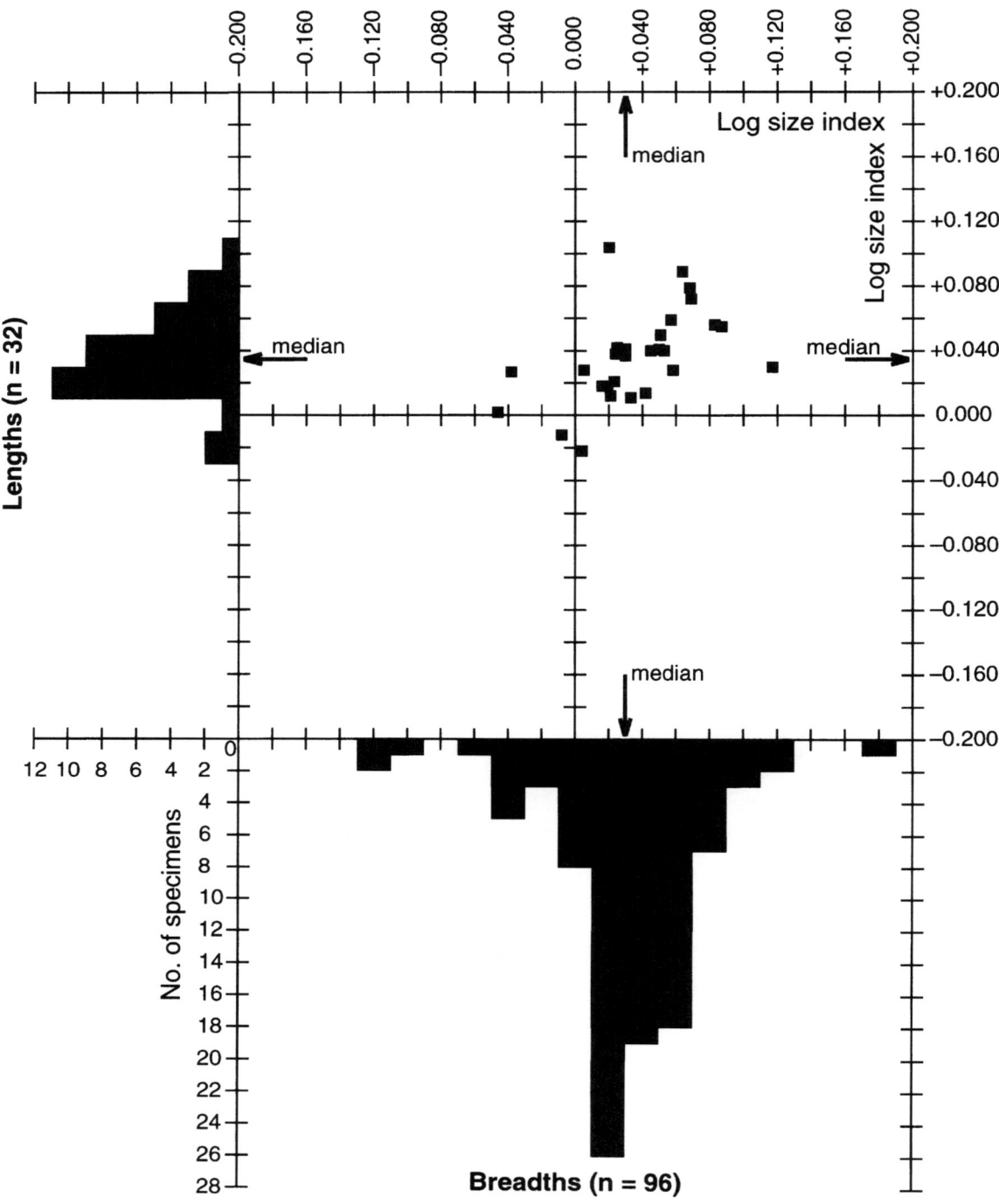

Fig. 8b. Çayönü *Sus*. Grill Building subphase.

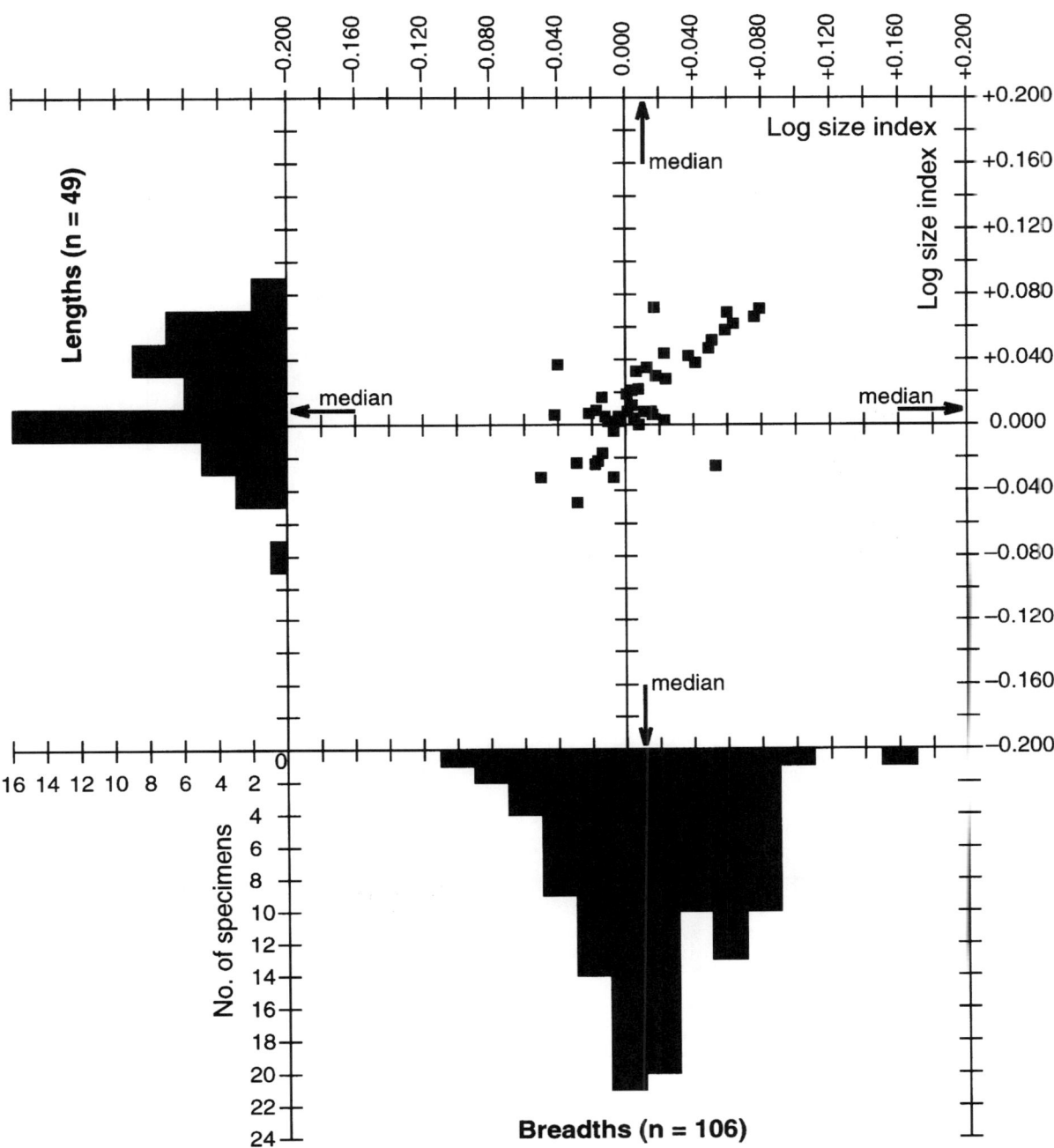

Fig. 8c. Çayönü *Sus.* Channeled Building subphase.

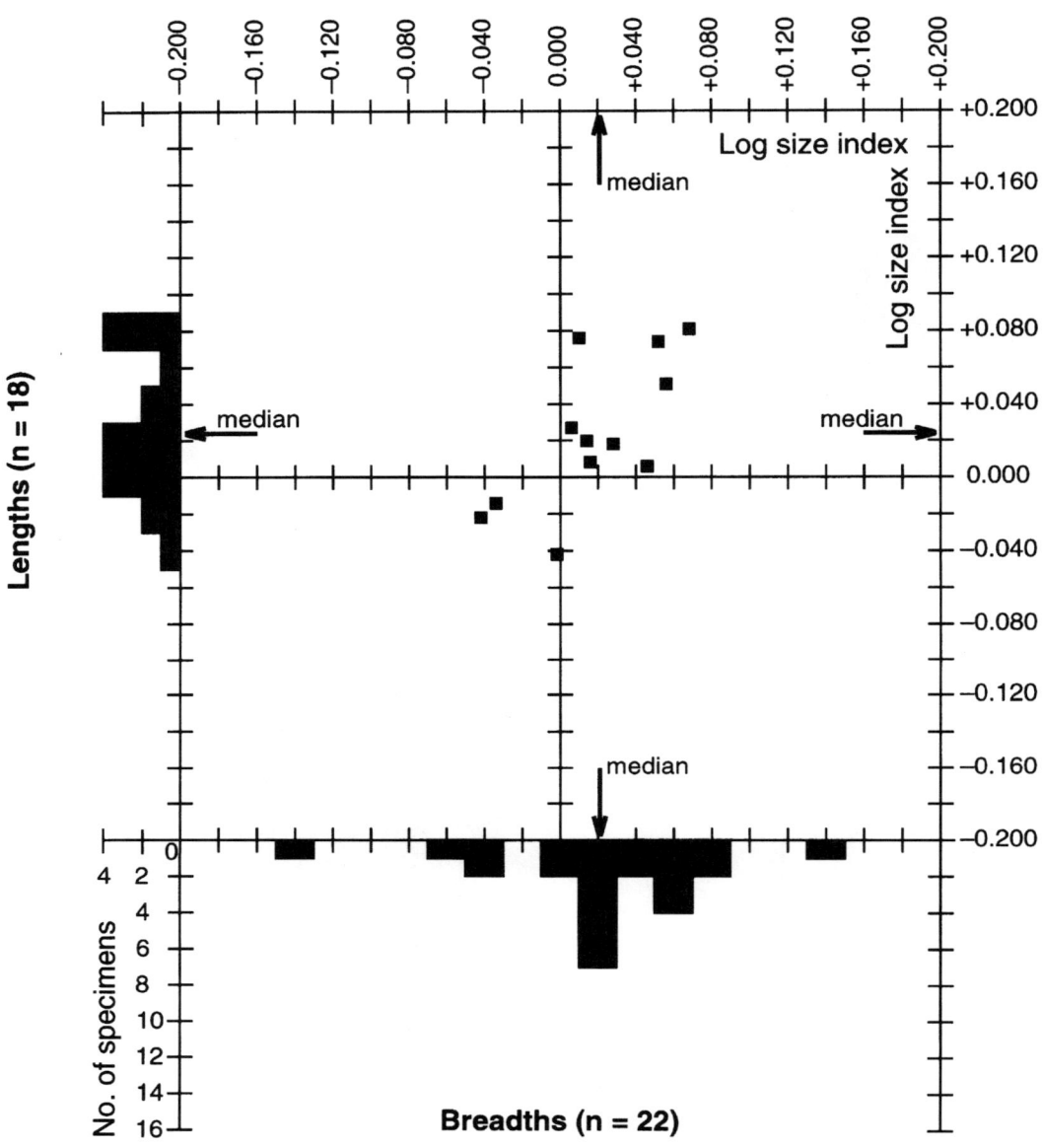

Fig. 8d. Çayönü *Sus.* Cobble-paved Building subphase.

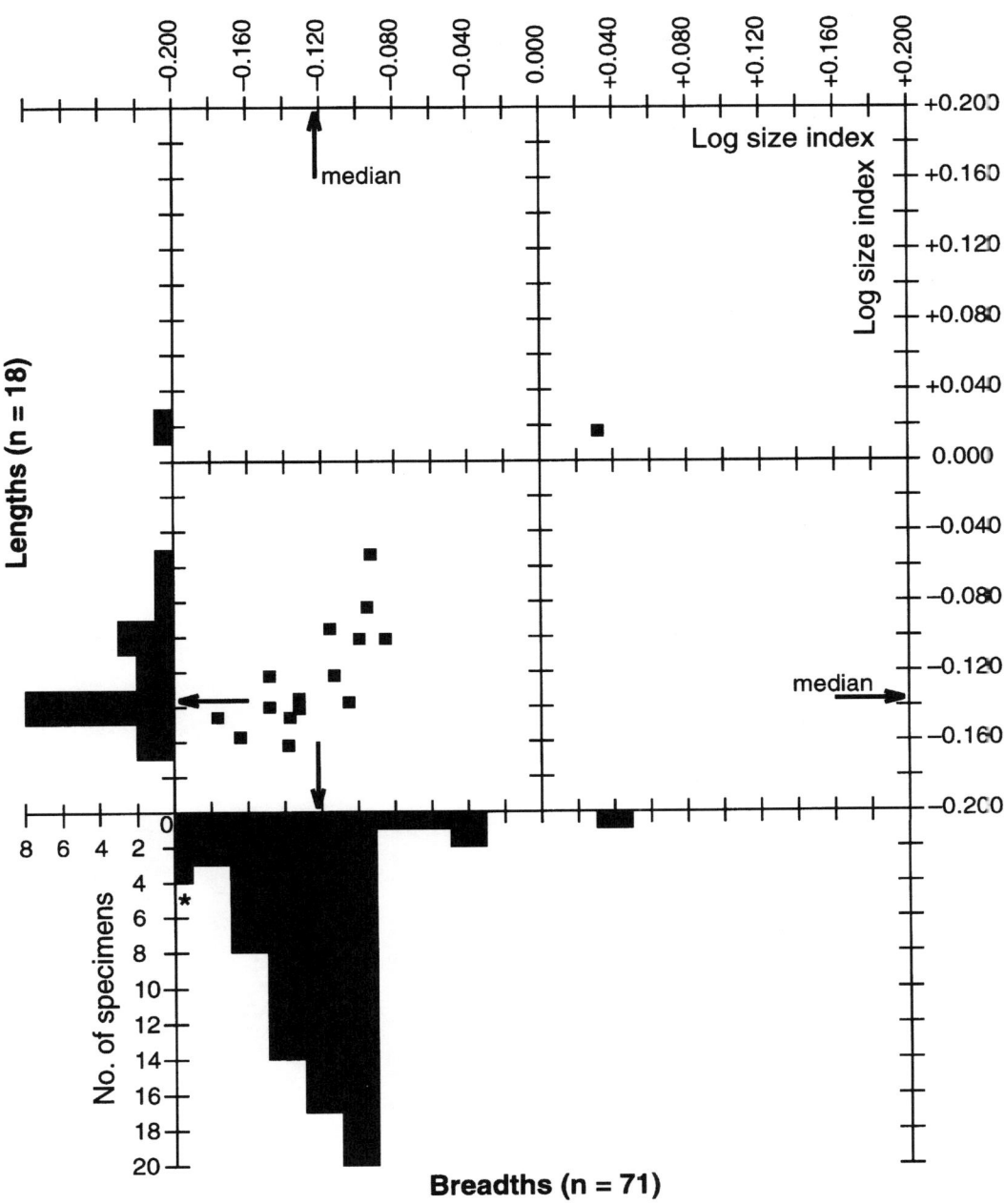

* = 4 specimens with LSI < –0.1901

Fig. 9. Kaman-Kalehöyük *Sus* log size index diagram. All temporal units combined. See the main caption to Figure 8.

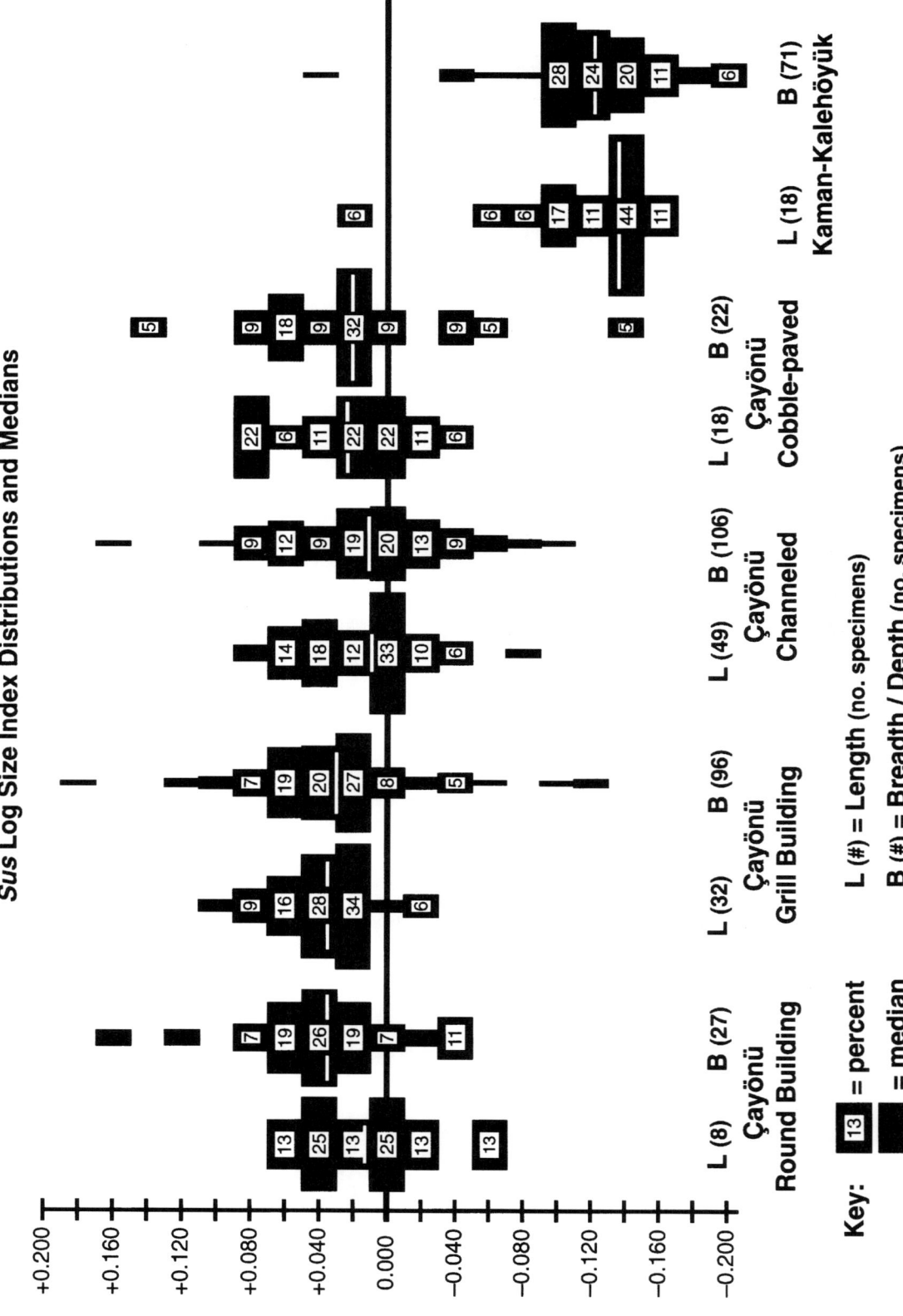

Fig. 10. Summary of Çayönü and Kaman-Kalehöyük *Sus* log size index data. The "battleship curves" show the percentage of indices falling into the various 0.020 intervals for lengths and breadth/depths for each temporal division. The median values are also plotted. The Round Building subphase is the earliest.

PIGS IN THE HONGSHAN CULTURE[1]

Sarah M. Nelson

Dept. of Anthropology, 2130 S. Race, University of Denver, Denver, CO 80208

Pigs are prominent in the ceremonial and ritual iconography of the Hongshan culture in northeastern China. While pigs have been important as food in most times and places in China since the Neolithic (Chang 1977; Anderson 1988), it is their ritual significance in the Hongshan culture that I will discuss in this chapter. Pig heads are carved on jades, parts of a life-sized statue of a pig were found in a ceremonial building, at least one burial contains pig bones, and a prominent mountain has a skyline that resembles the head of a pig. This combination of circumstances suggests an unusual importance of pigs in the Hongshan culture.

To situate the Hongshan culture in space and time, it is centered in western Liaoning province and eastern Inner Mongolia. It was first described by Japanese scholars (Hamada and Mizuno 1938) as a far northern offshoot of the Yangshao culture of the Yellow River region. The name Hongshan, meaning "red mountain," was given to the site and eventually to the culture because of the prominent red ridge that runs beside the town of Chifeng in Inner Mongolia. Little further attention was paid to this culture until it was discovered that the large stone-mounded tombs in the area were related to it. When these tombs were found to be the source of well-crafted small jades that had been appearing in art markets in Hong Kong and the West, serious efforts were made to find and record new sites. Through this effort the ritual centers of Niuheliang and Dongshanzui were discovered and partially excavated. Each is distinctive and unique. Niuheliang will be described more fully below, for it contains the life-sized pig statue and Pig Mountain. Dongshanzui is characterized by a number of low structures which are described as altars and walls. Neither site appears to be a habitation site (Guo and Zhang 1984).

These two ritual centers are more or less in the center of the area of known Hongshan sites (Fig. 1), although the densest concentration of villages is found to the north around Chifeng. The distribution of the Hongshan culture is around the Daling and Xiliao rivers, with outliers as far east as the Bohai coast (Andersson 1923), north of the Xiliao River valley, and south as far as the Yanshan mountains, where Hongshan abuts the Yangshao culture (Guo 1995). The Hongshan culture is dated to about 4000–2500 B.C. (Sun and Guo 1984). The "Goddess Temple" at Niuheliang, where the pig statue fragments were found, is specifically dated to 3500 B.C. by a piece of wooden post, with dates of 4970+80 B.P. and 4975+85 B.P., calibrated to 3625 B.C. and 3630 B.C. (Sun and Guo 1986a).

Among the indicators of pig ritual, most conspicuous are the *zhulong* (pig-dragons), slit annular ornaments with a pig head on one end of the ring (Fig. 2), found on or near the chests of skeletons in elaborate tombs in the central part of the Hongshan area. While these pieces are well represented in jade collections and hence widely known (Childs-Johnson 1988; Fang and Liu 1984; Huang 1988; So 1993), life-sized fragments of pig statues from the "Goddess Temple" in Niuheliang are not often described. Another suggestion of the importance of pigs is Zhushan (Pig Mountain), a ridge on which a prominent peak has the shape of a pig's head. This formation is particularly pig-like when viewed from the location of the Goddess Temple. Together these manifestations add up to an intense pig iconography which is not common in China (or elsewhere), and therefore demands explanation.

My questions regarding pigs in Hongshan iconography fall into two categories: Why pigs (rather than sheep or snakes, for example)?, and What (if anything) could pigs have had to do with the formation of complex society, especially social stratification and economic diversification? I believe the prominence of pigs is related to the organization of society, especially the formation of a class of elites, who were so conspicuously interred in the Hongshan mounded tombs. I invite the reader to make the leap from pig iconography to pig rituals, and to consider the plausibility of pig rituals as instrumental in the

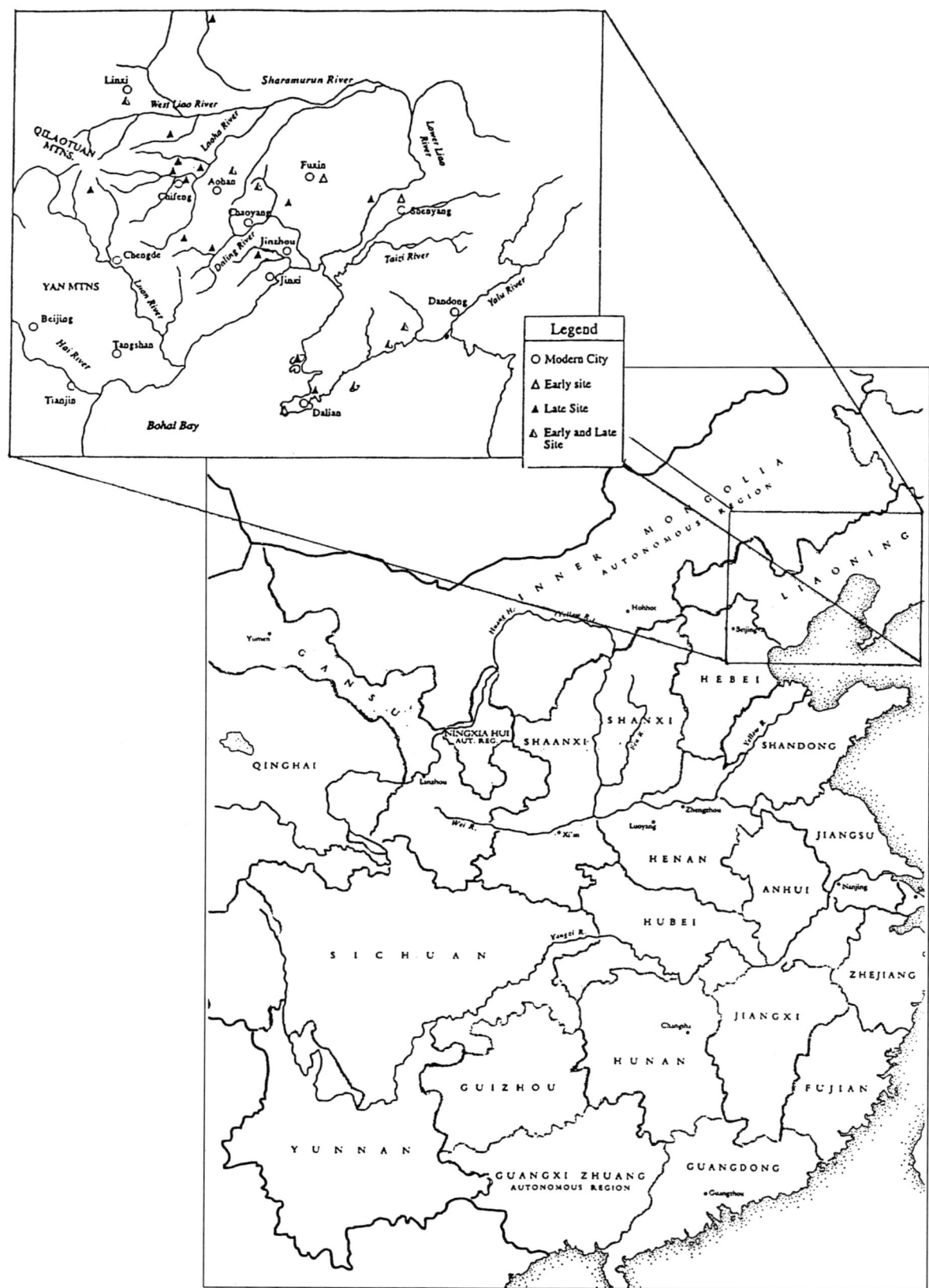

Fig. 1. Map of Hongshan sites in northeast China.

Fig. 2. Jade "pig-dragon."

creation of the elite level of society, the elite in turn using these rituals (among other things) to create and sustain their control over the populace. My argument requires two approaches: a consideration of the place of real pigs in the Hongshan economy on the one hand, and a detailed examination of the function of pigs in the ideological system on the other. The need for a managerial elite is suggested as both cause and effect of the pig manipulation.

Pigs in Asian Archaeology

Pigs are the earliest domesticated animal in China. Pigs were an early domesticate in China, whether the staple crop was rice or millet, and regardless of the local ecology. Pig bones, described as domesticated, are found in the site of Zengpiyan (between 9000 and 7000 B.C.), near Guilin in southern China (Chang 1986). No evidence of plant cultivation was found there, although cultivated rice is known from the site of Pengtoushan, somewhat farther north on the Yangzi River, by approximately 7000 B.C.; Richard MacNeish's data from southern China corroborate this finding (MacNeish and Libby 1995). In northern China, millets and pigs appear to have been domesticated at approximately the same time. The sites of the Peiligang culture, dated as early as 6500 B.C. in the Yellow River region, have ample evidence of rice cultivation, and abundant clay figurines of pigs were found at these sites (Chang 1986) (Fig. 3). Burials of whole pigs are also found occasionally. Millets were identified at Xinglongwa, in Aohan Banner, Inner Mongolia (IMAT 1985), a site with calibrated dates ranging from about 6000 to 4500 B.C. (Institute of Archaeology 1991), roughly contemporaneous with confirmed millet cultivation in central China.

In the Yangshao region on the Yellow River, pigs are found extensively but without evidence of ritual treatment. However, in the succeeding Longshan culture, pigs and parts of pigs become quite common in burials. It is not always a meat bone, but sometimes a skull or jaw that is included among the grave goods. A trade and feasting model has been used to explain pigs in Neolithic Chinese burials (Kim 1994), but it is one that seems to me not to fit the local conditions very well (Nelson 1994).

Pigs are present during the Shang Dynasty (ca. 1700–1100 B.C.), but they are mentioned in the oracle bones only in terms of hunting wild boar (Keightley 1978). In peripheral regions, some uninscribed oracle bones are made from pig scapulae (Chang 1986:270).

The Shang dynasty made a few bronze pig statues, but pig representations are rare. In Inner Mongolia, the Lower Xiajiadian culture features elite burials with whole pigs and dogs, and lesser burials with only parts of these animals (Guo 1995b). This tradition persists in the northeast, even as far north as Jilin province, into the middle of the first millennium B.C.

Down through time, as long as food offerings are found in Chinese burial contexts, pigs are among them, including dishes with pork. At Mawangdui, the completely preserved burial of a noble woman of the Chu kingdom contains meat offerings including pork. Ceramic figures of the Han Dynasty include pig sties for the future spiritual eating of the spiritual pigs by the spirits of the dead into eternity.

The pig bones found in shellmound sites are always assumed to be wild in South Korea, but they have not been studied with the question of domestication in mind. North Koreans report that domesticated pigs are found in early Neolithic sites, although this assertion lacks rigorous scientific reporting. The size of the pigs alone might be a clue, because the wild pigs of Korea are said to be enormous, up to 2 m long, while domestic pigs tend to be considerably smaller. The acorns that are occasionally found in Korean sites are always interpreted as human food (indeed a kind of acorn jelly is still made in Korea), but the possibility of acorns stored for pig consumption should be considered as well. Kim (1986) notes that Bronze Age statues of pigs had holes to accommodate real bristles, making them seem more lifelike.

Pigs for Food and Sacrifice in Northeast Asia

Pigs may be sacrificed in various situations, for a number of purposes. They may be offered in cooked dishes to the ancestors on appropriate occasions, but were infrequently ritually sacrificed among the Han Chinese. However, among the Manchu, pig sacrifice was the center of shamanistic ritual. In Korea, a cooked pig's head is often found on the altar of the *mudang*, usually a woman shaman, or a whole pig may be prominently displayed nearby. This object is used for divining purposes. If the pig can be balanced on a large trident, it is an important auspicious sign, evidence that the ceremony has been efficacious (Fig. 4). Pigs are symbols of wealth and prosperity in China.

Hongshan Archaeology

The Hongshan culture is deeply rooted in the Neolithic of northeastern China. The common brown pottery is clearly a development from earlier cultures (Nelson 1990), such as Xinglongwa (IMAT 1985), Chahai (Fang 1986), and Xinle (OPASPM 1985). The stone tools likewise demonstrate local continuity, and even the Hongshan jade-working echoes slit annular earrings from Chahai and an unfinished annular pendant from Zuojiashan (Jilin University 1989).

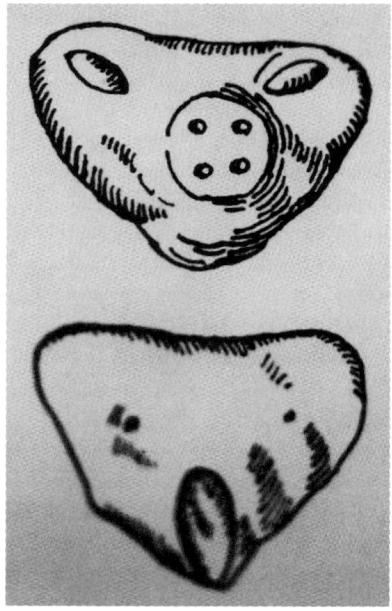

Fig. 3. Clay pig head from Peiligang.

Along with these familiar characteristics, significant new traits appear in Hongshan, including large mounded tombs, painted pottery, and realistic human and animal statues and human statuettes. More important than individual traits, however, are indications of complexity of many kinds. Monumental tombs imply social hierarchy, ceremonial architecture demonstrates a rich ritual life, and the presence of (presumably symbolic) emblems buried with the dead suggests a total society too large for every important individual to be recognized on sight, or one in which a person's rank was more important than his or her individual achievements.

Complexity is implied in the economic system, as well. Although the total economic system cannot be known from the present state of excavation and reporting, some aspects can be inferred. The likelihood of craft specialization is implied by the skill shown in the crafts of painted pottery and jade carving, as well as in the number of goods produced. The quantity of painted cylinders in particular requires organization of production and distribution by a managerial elite. The presence of at least two large ceremonial sites (Niuheliang and Dongshanzui) demonstrates organization and building beyond the village level. Thus the archaeology of Hongshan reveals a society in which sites, buildings, graves, and artifacts are organized in new ways not known earlier in China.

The earliest reported sites of the Hongshan culture were villages with semi-subterranean dwellings. These sites contained both brown pottery vessels decorated with zigzags and other surface impressions, such as were found throughout the northeast in the Neolithic, and black-on-red painted pottery which was a new addition. The excavators of Hongshanhou, which became the type site, thought the painted styles were similar to those of Gansu province (Hamada and Mizuno 1938). Stone tools were more numerous than potsherds, and included both chipped and polished artifacts. Large and small axes, large hammerstones, knives, awls, scrapers, arrowheads, hoes, and stone plows were among the stone tools constructed according to specific templates, along with grinding stones, pestles, whetstones, and polishing stones. Spindle whorls made of potsherds were presumably recycled from broken pots, and marine shells provided raw material for bracelets. Faunal remains reported from the first excavation included deer, pig, sheep, cattle, and possibly horse (Guo 1995). (The horse bones, however, are only from the Bronze Age layers.) These ordinary characteristics made Hongshan at first appear to be just another Neolithic culture, a bit far into barbarian territory north of the Great Wall, to be sure, but classifiable as a northern off-shoot of the Yangshao culture.

Even before discovery of the ceremonial centers, however, when only the village sites were known, a certain amount of task complexity could be inferred from the number of different tool types and their standardization. The variety of tools implies that many different activities were carried on in the villages: plant tillage, animal husbandry, cloth and leather production, and the manufacture of the stone tools themselves. One of the tools, the stone *si*, is considered by some archaeologists to be a plow (Sun and Guo 1984), the earliest local indication of the intensification of agriculture.

Pigs in Neolithic Sites in the Northeast

Although pig bones are found commonly in Neolithic sites, outside of Hongshan there are no pig images other than small figurines in the Neolithic. By the beginning of the Hongshan culture pigs had probably been domesticated in this region for two millennia. They were thoroughly familiar and must have been an important part of the subsistence base. An agricultural economy had long been established in the preceding cultures of Liaoning and Inner Mongolia. The brown pottery with rocker-stamping over the exterior is similar to that of other Neolithic cultures in the north, from the Peiligang and Cishan cultures in the Central Plain to Zuojiashan in Jilin province. A prominent characteristic of some of these sites is the presence of ceramic pig figurines (Chang 1986).

In spite of their frequency, the pig figurines cannot reasonably be interpreted as objects of worship in those early contexts. They were found in house pits and trash pits, and many were made rather casually. Other animal figurines found at various sites include birds, lizards, and monkeys, to name a few (Yang 1988). But pig figurines are by far the most common, suggesting an uncommon importance of pigs.

Pigs might have merited attention in the form of figurines for a number of reasons. Perhaps pig health was important to the prosperity of the family, or perhaps pigs represented prosperity in themselves (as a measure of exchange value, for example), or perhaps whatever spirits existed liked pork, but being non-corporeal, they were satisfied with clay pigs. Wolfram Eberhard (1968:260) notes that pigs were the animal most important in sacrifice.

Quantified data on pigs would be useful—their relative and absolute frequencies at various sites, their ages at slaughter, butchering techniques, and so forth. This type of information unfortunately is rarely either systematically collected or reported in Chinese archaeology. Lacking those important details, we may at least determine that, whatever the function of northeastern Chinese pig figurines, they probably indicate the importance of pigs in the farming economy. Consider the benefits of pigs. Pigs in temperate forest climates scavenge for themselves, so that no effort by humans is required to provide them with food. Some of the foods they scavenge, such as acorns, are edible by humans only with tedious processing, while pigs process it easily into edible protein. Furthermore, pigs help to keep the village tidy and cut down odors by eating food scraps and human waste (Nemeth 1995; Nelson 1994; Eberhard 1968:304). Pigs also produce fertilizer which can be used to enhance crops. Pigs, like dogs, imprint upon their human owners, know where they live, and are unlikely to run away. Perhaps, like the pig-keepers of Oceania (e.g., Rappaport 1967), the Neolithic farmers did not eat meat regularly, but used pigs as a supplement to their diet in a world punctuated by episodes of feasting, as has been suggested for other Chinese sites (Kim 1994). Inedible parts of the pig were surely processed into tools, containers, and clothing in the same ways that wild animals would have been. Every family probably had a pig or two to keep them stocked with meat on the hoof, as well as with raw material for awls, cloaks, and bags,

Fig. 4. Korean *mudang* (shaman) dancing with cooked pig head at a *kut* (ceremony).

and to serve as a living vacuum cleaner.

In dry climates, a major drawback to pigs is that they require moisture on their skins. During any extended arid period, the pigs would suffer and would require extra care. What would happen to pigs if desertification were beginning at the eastern edge of the Mongolian grasslands, in what is now western Liaoning and eastern Inner Mongolia? In the changing ecosystem, pigs would become increasingly difficult to raise, and as a consequence they would probably become increasingly scarce. Desiccation of the environment of eastern Inner Mongolia has been recognized in geomorphological studies (Wang et al. 1992), but the timing of the onset of aridity has not been precisely determined. Evidence for increasing aridity in Liaoning appears to begin at 3000 B.C. (LQPLR 1978), but this estimate is based on uncalibrated dates. Therefore, the arid period probably coincides with the ceremonial centers at Dongshanzui and Niuheliang. This association needs to be confirmed with further studies.

The probable association of pig ceremonialism and the onset of increasing aridity raises the possibility that a subset of the population could have become elite by acquiring control of pigs. The suggestion here is that pigs were valuable and becoming a scarce commodity. It is easy to produce some scenarios—private access to a swampy place, an artificial pond, a strain of pigs that required less water, for example—but none of these is extant in the archaeological record. The only certainty is that during the Late Hongshan period, when the ceremonial centers were at their height, real pigs still existed, but apparently not in large numbers. Judging from the existing reports of faunal remains in ritual contexts during the course of the Hongshan culture, sheep became more common than pigs. For example, pit H1 at Niuheliang, south of the Goddess Temple, contains mainly sheep bones. Pig and ox bones are mentioned as present in the tombs only once (Fang and Wei 1986b). It is suggested that as sheep began to replace pigs in the agricultural economy, a nostalgia for pigs, and perhaps a belief that the ancestors' spirits required pigs for sacrifices, could have given special power to the owners of the dwindling supply of pigs.

Whether the farmers were willing to cede some of their autonomy for occasional access to pigs is difficult to assess, but it is clear that pig iconography became increasingly sophisticated and important with the rise of the elite. These pigs were not memorialized with small, crude pottery figures, but were created in rare jade by skilled craftsmen. These jades probably indicate the highest class, since the *zhulong* are represented in the round while other pendants are flat with bas relief carving. The locus with a prime view of Zhushan, the pig-headed mountain mentioned earlier, became a center of ritual activities at Niuheliang. The temple that was built at this sacred location contained, among other statues, at least one life-sized pig statue which must have been convincingly life-like, to judge from its remaining fragments. Perhaps the pig had even become an oracular figure. The statues of at least one bird and several female statues of different sizes were also found in the temple. The temple building was large and complex, with additional rooms behind the main room which held the statues (Sun and Guo 1984, 1986a).

Whatever rituals occurred at the temple, it seems that designated officiants must have been in charge of those activities. Perhaps, based on the fact that the human statues depict women, a priesthood of women may have managed the temple complex (Nelson 1991). It may be a stretch to associate the sacred pigs with women, although in Manchu times women were the keepers of pigs (Lattimore 1951).

Some individuals, whoever they were, acquired or inherited the right to wear an emblem of the deified or spiritual pig, no longer with an earthly and realistic body, but now a pig head on a perfect circle, without legs or tail. Even the head, although still recognizable as a pig, had become stylized, with enormous round eyes. Were the individuals who were entitled to wear these pendants the keepers of the pigs and/or the keepers of the temple?

The *zhulong* appear to be the most important of the jade symbols, as these pendants are fully three-dimensional, while the others are flat and usually one-sided. The flat jade pendants include turtles, owls, flying birds, and clouds, emphasizing air and water. Of these, the cloud emblems are the most elaborately made, and for this reason they seem likely to belong to those second in rank, perhaps having a special role related to water usage. In an increasingly arid area, water would have assumed an ever more urgent importance. Turtles, who live in water, imply a similar meaning. It is interesting to note that Eberhard (1968:259–260) links pigs with thunder in the person of a god who regulates rivers.

Control of Production

To whatever extent the elite gained control of the economy, they accomplished this feat by persuasion or by providing services, for there is no evidence of force of any kind. In addition to the possibility of control over pigs, purely elite needs also arose. The emerging elite required status markers to set themselves apart from the farmers and artisans. Hundreds of villages supported the ceremonial centers, and their inhabitants would not have recognized elite individuals by sight. Therefore, visible symbols would have been necessary to designate those persons who commanded respect or obedience.

Status markers developed in at least four new ways: in production of emblematic jade objects, in ritual pottery, in building the large tombs and other ritual constructions, and in the products of a fledgling metal industry (Han 1993; Guo 1995). All of these activities required training of skilled artisans and management of production. For jade and metal, the need for non-local raw material would have added the dimension of long-distance procurement, or perhaps trade with distant peoples. Marine shells, although not common, by their very presence constitute an echo of long-distance trade.

While the locations where objects for elites were produced are not specifically identified, there is no evidence of such production in excavated villages. Pottery for daily use, stone tools, and cloth continued to be produced at the small sites. Kilns have been excavated at Xilingshan in Aohan Banner, Inner Mongolia (Guo 1995), showing that some pottery production occurred at the village level. Furthermore, no evidence of storage of agricultural produce has been found at the central sites, suggesting that the elites had not acquired control over food production or distribution.

Were the ritual centers the locus of production for the elite? Adjacent to the Goddess Temple at Niuheliang, just above it on the hill slope, a large platform was constructed, edged with stones (Fang and Wei 1986b). The function of this platform is unconfirmed, but a possible indication of its use is the presence of sherds of large tubular painted vessels with thin walls. These bottomless pots are found almost exclusively in elite tombs, and sherds from these vessels have also been found in the Goddess Temple. Each elite tomb contained several rows of these pots, arranged in a square around the central burial. The vessels, 30–50 cm wide, were placed abutting each other in a line. It has been calculated that the burials at Niuheliang alone must have contained about 10,000 of these pots (Guo 1995), a staggering amount of pottery production, even though the need for them would have been episodic, arising only with the death of a major leader.

It is reasonable to suppose that these large and fragile vessels were manufactured near the place of their use to avoid the problem of transporting them over great distances, presumably one or two carried by each human porter. While the pots are not identical, they are uniform enough to suggest centralized direction. Steps involved in pottery production at this scale would have required skilled artisans, especially for shaping, firing, and painting the vessels, as well as other workers to procure the clay and paint supplies, prepare the clay, tend the fires, and so forth. This complex process would have required managers, given the scale of production indicated by the tombs and the fact that each tomb required large numbers of vessels as needed.

No definitive site of jade production has been discovered at Niuheliang, although some jade chips have been noted which may indicate a production area (Guo 1995). Elite burials contain up to 15 jade objects per individual buried, an astounding number in the face of a considerable amount of tomb robbing and trafficking in these artifacts for more than one hundred years. Hundreds, perhaps thousands, of these jades have found their way to museums and collectors around the world. These jades are crafted in specific forms, as noted above. Although fewer jades were expended per burial than tubular pots, nevertheless a great many must have been carved. Three points are critical in viewing these artifacts as evidence of managed production. First, the artisans were not free to create any shapes they fancied, but were constrained to carve specific shapes and sizes. Second, the jade is a non-local raw material, requiring some means of transportation to the site. Finally, a continuing stream of new jades had to be crafted because the old ones were constantly going out of circulation upon the death of their owners. These characteristics combined suggest that control of production by a managerial elite was necessary.

Another salient feature at Niuheliang and elsewhere is the enigmatic large ritual constructions which would have required the mobilization of considerable labor power, as well as an individual or team to design the construction and direct its completion. These features include platforms, walls, and at least one temple and one earth pyramid, as well as numerous complexes of large tombs. Construction of the tombs would have been episodic, but each would have required enormous labor power. The structure of the tombs was elaborate, sometimes with a central well-made square cairn, an outer square stone structure, and subsidiary burials of elites of lesser status in stone-slab coffins ranged around the central burial.

It is relevant to note that these subsidiary burials include both primary interments of a whole body and bundles of bones implying secondary interments (Guo 1995). It seems likely that bodies of individuals who were selected to accompany the central figure to the afterworld, but who predeceased that individual, were kept in a sacred place awaiting burial. That place might well have been the back rooms of the Goddess Temple (J1A), or perhaps the nearby building (J1B) (Liaoning Province 1986). Wherever the bodies were kept, an elaborate mortuary ritual, requiring many hours of labor under supervised direction, is implied by the archaeological discoveries.

Finally, and most controversially, is the possibility of metal production. Little metal has been found, and the dates of the sites are quite early for copper or bronze in China, although contemporaneous with some Qijia bronze. An unprovenienced clay mold has been reported (Anonymous 1988), and an earring with a copper wire loop was found in situ in a recently excavated Hongshan grave (Han 1993). Most provocative of all is an unpublished find on the top of a pyramid-like construction, with stepped rings of vertical stone walls within the earthen matrix. Crude potsherds, interpreted as crucible fragments, were found near the top of the pyramid. These circumstances suggest that metal casting was considered sacred, perhaps magical, to be conducted on a high platform in full view of the populace. The ability to turn ore into metal tools or ornaments might well have been associated with supernatural powers. The evidence for this reconstruction is thin, but tantalizing.

The sources of the raw jade used for carving the emblems of Hongshan are not known, but most of them are not local. Other than Xiuyan jade from the hills near Dandong, Liaoning province, sources of jade have not been reported. Xiuyan jade is uniformly light green, while the jade objects from the Hongshan culture are found in many colors. Even turquoise, known to occur naturally only in Central Asia, appears among the objects. Thus, the various colored stones, acquired from distant locations, had to be either quarried on long journeys or received in trade.

If trade rather than simple quarrying was the mechanism of acquiring jade, it is interesting to consider what the people of Hongshan traded for the attractive stone. The presence of spindle whorls bespeaks cloth production, and numerous sheep bones show that wool could have been woven. Cloth is relatively easy to transport, and was used as a medium of exchange in many ancient economies. There is, however, no indication that cloth production was under the control of the elite. For example, no concentrations of spindle whorls or loom weights have been found. However, spinning might well have been accomplished by individuals in villages, or village production of cloth could have been partially taken in taxes by the elite, or offered in tribute to the temple. Agricultural products seem less likely trade objects because they are bulky to transport over long distances.

Conclusion

Whatever the mechanism for acquiring the jade stone, and whether or not metal played any part in the economy of Hongshan in the middle of the fourth millennium B.C., it is clear that an elite had set up and gained control of the production of certain goods that were for their use alone. These include, at a minimum, emblematic jades, painted pottery cylinders, ritual constructions, and realistic statues ranging in size from a few centimeters high to life-sized and larger. The elite seem to have comprised a managerial class, perhaps producing nothing themselves. The ability to expend without being producers demonstrates a fundamental split between workers and elite.

There is no evidence of either warfare or coercion of the local populace in Hongshan sites. It is likely that the elite formed gradually, and obtained power by claims to special knowledge of the supernatural (see also Nelson 1996).

Probably production of objects for rituals was glossed as working in the service of the supernatural, not in the service of the elite. Thus a managerial class arose, in the guise of filling a need for serving the gods, and at the same time managing the pigs for the increasing prosperity of the whole society. Pigs, symbols of prosperity, may have played an important role in the construction of that elite.

Note
1. This chapter is revised from Nelson 1995 for this volume.

References

Anderson, E.N. 1988. *Food of China*. Yale University Press, New Haven.

Andersson, J.G. 1923. The Cave Deposit at Shaguotun in Jingxi, Fengtian. In *Palaeontologia Sinica*, ser. D, 1, Vol. 1, whole issue. Geological Survey Institute.

Anonymous. 1988. Another Important Discovery of the Chinese Northern Cultural Relics—Neolithic Casting Mold Unearthed in Inner Mongolia. *Renmin Ribao* (Jan. 10): 3.

Chang, K.C. 1977. *Food in Chinese Culture*. Yale University Press, New Haven.

——— 1986. *The Archaeology of Ancient China*. Yale University Press, New Haven.

Childs-Johnson, Elizabeth. 1988. Dragons, Masks, Axes, and Blades from Four Newly-documented Jade-producing Cultures of Ancient China. *Orientations* (April): 30–37.

Eberhard, Wolfram. 1968. *The Local Cultures of South and East Asia*. E.J. Brill, Leiden.

Fang Dianchen and Wei Fan. 1986a. Excavating a Lost Culture. *China Reconstructs* (Dec.): 33–39.

——— 1986b. Brief Report of the Excavation of Goddess Temple and Stone Graves of Hongshan Culture at Niuheliang in Liaoning Province. *Liaohai Wenwu Xuegan* 8:1–17. In Chinese.

Fang Dian-Chun. 1986. A Brief Report on Exploratory Excavation of Neolithic Ruins in Fuxin Cha-hai. *Liaohai Wenwu Xuegan* 8:1–17. In Chinese.

Fang Dian-chun and Liu Bao-hua. 1984. Discovery of the Hongshan Culture Jade Tombs at Hutougou of the

Buxing County in Liaoning. *Wenwu* 6:1–5. In Chinese.

Guo Dashun. 1995a. Hongshan and Related Cultures. In *The Archaeology of Northeast China: Beyond the Great Wall*, ed. S.M. Nelson, pp. 21–64. Routledge, London.

——— 1995b. Lower Xiajiadian Culture. In *The Archaeology of Northeast China Beyond the Great Wall*, ed. S.M. Nelson, pp. 147–181. Routledge, London.

Guo Dashun and Zhang Keju. 1984. Brief Report on the Excavation of Construction Sites of Hongshan Culture at Dongshanzui in Kezuo County, Liaoning Province. *Wenwu* 11:1–11. In Chinese.

Hamada Kosaku and Mizuno Seiichi. 1938. *Hung-shanhou, Chihfeng. Prehistoric Sites at Hung-shan, Chihfeng, in the Province of Jehol, Manchukuo*. Archaeologia Orientalis, Series A, Vol. 6. In Japanese with English summary.

Han Rubin. 1993. Recent Archeological Metallurgical Achievements at the University of Science and Technology, Beijing. Paper presented at the international conference Chinese Archaeology Enters the 21st Century, May 29.

Huang Xuanpei. 1988. China's Neolithic Jadeware. In *Ritual and Power: Jades of Ancient China*, Elizabeth Childs-Johnson, curator. China Institute in America, New York.

IMAT (Inner Mongolia Archaeological Team). 1985. Excavation of the Xinglongwa Neolithic Site, Aohan Banner, Inner Mongolia. *Kaogu* 10:865–874. In Chinese.

Institute of Archaeology. 1991. *CASS, Radiocarbon Dates in Chinese Archaeology, 1965–1991*. Cultural Relics Publishing House, Beijing.

Jilin University Archaeology Teaching and Research Section. 1989. The Neolithic Site at Zuojiashan, Nong'an, Jilin Province. *Kaogu Xuebao* 2:187–212. In Chinese.

Keightley, David. 1978. *Sources of Shang History*. University of California Press, Berkeley.

Kim, Seong-og. 1994. Burials, Pigs, and Political Prestige in Neolithic China. *Current Anthropology* 35:199–241.

Kim Wonyong. 1986. *Art and Archaeology of Ancient Korea*. Taekwang Publishing Company, Seoul.

LQPLR (Laboratory of Quaternary Palynology and Laboratory of Radiocarbon). 1978. Kweiyang Institute of Geochemistry, Academia Sinica "Development of Natural Environment in the Southern Part of Liaoning Province During the Last 10,000 Years." *Scientia Sinica* 4:516–532.

Lattimore, Owen. 1951. *Inner Asian Frontiers of China*. American Geographical Society, New York.

Liaoning Province Cultural Relics and Archaeology Institute. 1986. Brief Report on the Excavation of the "Goddess Temple" and the Stone Graves of the Hongshan Culture at Niuheliang in Liaoning Province. *Wenwu* 8:1–17. In Chinese.

MacNeish, Richard S., and Jane G. Libby (eds.). 1995. Origins of Rice Agriculture. The Preliminary Report of the Sino-American Jaingxi (PRC) Project, SAJOR.

Nelson, Sarah M. 1990. The Neolithic of Northeastern China and Korea. *Antiquity* 64:234–248.

——— 1991. The Goddess Temple and the Status of Women at Niuheliang, China. In *The Archaeology of Gender*, ed. D. Walde and N. Willows, pp. 302–308. Proceedings of the 22nd Annual Chacmool Conference, Calgary, Alberta, Canada.

——— 1994. Comment on "Burials, Pigs and Political Prestige." *Current Anthropology* 35(2): 135–136.

——— 1995. Ritualized Pigs and the Origins of Complex Society: Hypotheses Regarding the Hongshan Culture. *Early China* 20:1–16.

——— 1996. Ideology and the Formation of an Early State in Northeast China. In *Ideology and the Early State*, ed. H. Claessen and J. Oosten, pp. 153–169. E.J. Brill, Leiden.

Nemeth, David J. 1995. On Pigs in Subsistence Agriculture. *Current Anthropology* 36(2): 292–293.

OPASPM (Office for the Preservation of Antiquities and Shenyang Palace Museum). 1985. The Second Excavation of the Neolithic Site at Xinle in Shenyang. *Liaohai Wenwu Xuegan* 2:187–212. In Chinese.

Rappaport, Roy. 1967. *Pigs for the Ancestors: Ritual in the Ecology of a New Guinea People*. Yale University Press, New Haven.

So, Jenny. 1993. A Hongshan Jade Pendant in the Freer Gallery of Art. *Orientations* (March): 87–92.

Sun Shoudao and Guo Dashun. 1984. On the Primitive Civilization of the Liao River Basin and the Origin of Dragons. *Wenwu* 6:11–20. In Chinese.

——— 1986a. Discovery and Study of the "Goddess Head Sculpture" of the Hongshan Culture at Niuheliang. *Wenwu* 8:18–24. In Chinese.

——— 1986b. Hongshan: A Lost Culture. *China Pictorial* 8:2–7.

Wang Jingai, Shi Peijun, Guo Suxin, and Suo Xiufeng. 1992. Research into Environmental Archaeology in the Great Wall Belt of Inner Mongolia in China. Paper delivered at the International Academic Conference of Archaeological Cultures of the Northern Chinese Ancient Nations, Hohhot, Inner Mongolia, August.

Yang Xiaoneng. 1988. *Sculpture of Prehistoric China*. Tai Dao Publishing, Hong Kong. In Chinese with English summary.

PIGS AND EMERGENT COMPLEXITY IN THE ANCIENT NEAR EAST

Melinda A. Zeder

Department of Anthropology, National Museum of Natural History, Smithsonian Institution, Washington, DC 20560

Introduction

Pigs (*Sus scrofa*) are arguably the least well understood but most intriguing component of ancient Near Eastern animal economies. Wild pigs are indigenous and plentiful throughout the region. They were an early (perhaps the earliest) species to undergo domestication (see Redding and Rosenberg, this volume). They can out-produce other livestock animals in numbers of offspring and meat per individual, and their meat has the highest fat and caloric content of any domesticate. And yet not only did pigs fail to become leading players in Near Eastern subsistence economies, they eventually became a reviled, pariah species for the vast majority of Near Eastern peoples. The reasons for the failure of this highly productive and widespread species to play a greater role in subsistence economies in the ancient Near East has been the focus of a great deal of discussion and debate by archaeologists, anthropologists, biologists, historians, and even political scientists. Using data from sites in the Levant and Mesopotamia, this paper explores some of the primary theories offered to explain why pigs failed to live up to their productive potential in this region, in the hopes of arriving at some new understandings of the place of pigs in emergent complexity in the ancient Near East.

Pig Keeping in Arid Environments: Pluses and Minuses

Before considering a number of case examples of human use of this species, let us first examine some of the basic pluses and minuses involved in keeping pigs in the Near East. There are a number of special constraints to raising pigs in arid environments. Several of these constraints stem from basic physiological attributes of the species. Particularly problematic is the pig's inability to convert high-cellulose plants, like grasses, into edible pro-

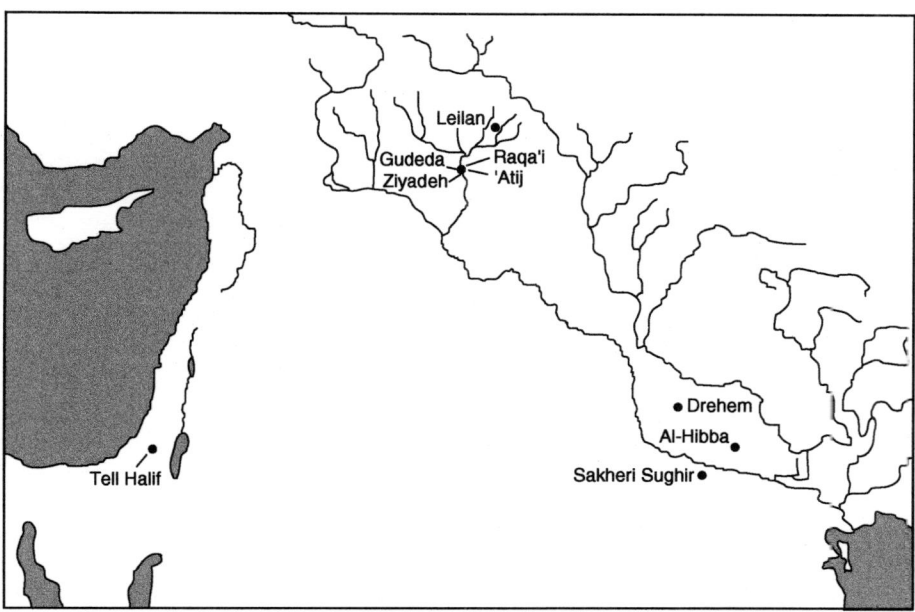

Fig. 1. Map showing location of sites mentioned in the text.

Table 1. Tell Halif periods of occupation and site functions

Period/Stratum	Dates	Function
Iron Str. VII–VI	1200–700 B.C.	Fortified Town
Late Bronze Str. VIII–XI	1550–1200 B.C.	Trade Outpost
Middle Bronze	1850–1550 B.C.	Hiatus
Early Bronze IV	2300–1850 B.C.	Hiatus
Early Bronze IIIB Str. XII–XIV	2500–2300 B.C.	Unfortified Town
Early Bronze IIIA Str. XV	2600–2500 B.C.	Fortified Town
Early Bronze II	2900–2600 B.C.	Hiatus
Early Bronze I Str. XVI–XVII	3200–2900 B.C.	Diffuse Settlement

tein. Since cereal grasses are the primary pasture plants in the region, pigs are at a definite competitive disadvantage in this regard compared to other livestock species (Harris 1974:34). Pigs also have higher water requirements than either sheep or goats or even cattle, and must be kept close to ample water supplies (Van Loon 1978:94). Lacking sweat glands, pigs also have a lower heat tolerance than other livestock species in the region, where summer temperatures regularly reach 100° F or more (Harris 1974:35–36). The range of management strategies that can be used to raise pigs in the arid Near East is thus limited.

Although wild pigs are known to move seasonally between river bottoms and hillside oak parkland (Hatt 1959:14–16), and despite some impressive examples of long-distance movement of domestic pigs during historic periods (see Perry 1985:91), the physiological attributes mentioned above and the species' legendary obstinate temperament all mitigate against the kind of long-distance movement that has been so successfully used in the management of sheep and goats in the Near East (Van Loon 1978:20). In addition, keeping pigs close to home carries its own special problems. Although pigs cannot utilize cellulose-rich grasses as effectively as bovids, they are known to trample and uproot cereal grains if allowed to venture into cultivated fields (Redding 1991). As a result, keeping free-ranging pigs close to sedentary settlements can wreak havoc with gardens and agricultural fields. Nor does the large-scale sty-based pig production seen in the farm belt of the United States seem to be a viable alternative. Pigs' higher water requirements, lower heat tolerance, and their feisty, quarrelsome nature (that comes to the fore under close conditions) would make this kind of management system a highly labor-intensive occupation in the absence of modern technology.

Finally, pigs are also less attractive than other Near Eastern livestock in the range of products they provide. Most other domesticated species in the region supply both non-renewable resources (meat, skin, and bone) and renewable resources (milk, wool, hair, labor). Pigs, however, are limited to non-renewable resources, primarily meat.

If raised under the right conditions, however, there are quite a few pluses to keeping pigs in the Near East. Although pigs cannot utilize pasture grasses as effectively as bovids, they can utilize a large variety of foods, including seeds, fruits, nuts, eggs, larvae, carrion, garden clippings, rotted produce, household garbage, and even partially digested food in the feces of other animals. Through this broad-base omnivory pigs are able to recycle waste and spoilage and convert it into a nutrient-rich food resource (Zeuner 1963:262–263). And while pigs have lower heat tolerance and higher water requirements than other livestock species, these limitations can be compensated for by keeping pigs close to home and providing them with shade and wallow (Van Loon 1978:72–73; Mount 1968). Also, while pigs may not be suited to long-distance transhumance, to free-range pasturing near fields, or even to large-scale sty management in arid environments, they are well suited to small-scale sty management where they can be fed household scraps. Keeping pigs this way requires little labor input, while yielding highly productive returns. And here is where pigs really shine. Though it is true that pigs do not provide the variety of resources of other livestock species, they out-compete all other domesticated animals in breeding frequency, number of young per litter, rate of growth to harvestable size, and in the quantity and caloric value of their meat yield (Van Loon 1978:26).

Theories of Pig Avoidance

Reasons offered for the paucity and eventual prohibition of pig in this region have ranged from functionalist/adaptationalist theories that emphasize the health risks of eating undercooked pork, to psycho-structural theories which look to various physiological or behavioral attributes that make the pig a reviled or symbolically dangerous animal (cf. Douglas 1966). For the purposes of this paper, I will focus on more "materialist" theories of pig avoidance that concentrate on those features of physiology or behavior that may limit the animal's suitability to Near Eastern environments and pastoral strategies. One of the most prominent of the theories emphasizing the physiological limitations of pig raising is that developed by Grigson (1987) for pig avoidance in the Levant. Grigson maintains that their high water requirements and low heat tolerance are the primary limiting factors to keeping swine in arid environments. She predicts that the proportion of pig in the diet will vary inversely with increasing aridity in the region, and uses the 200 mm per year rainfall isoheyet as the primary cut-off point for keeping pigs in the Near East.

Krader (1955), Zeuner (1963:260), and Flannery (1983), in contrast, have all linked pig avoidance to the fact that these animals are ill-suited to nomadic pastoralism. Long-distance transhumance has been used for millennia in this region to take full advantage of pasture resources and to limit conflicts between agriculture and pastoral production. Moreover, given the political uncertainties of settled life endemic to the region, sedentary populations might also be more inclined to keep animals that can easily accommodate a sudden shift to a more mobile lifestyle. Sheep and goats provide their keepers with this kind of flexibility, while the physiological requirements and basically truculent nature of pigs provide herders with only limited management options and, thus, preclude them from being major players in Near Eastern animal economies.

Building on Harris's theories (1974, 1985) that posit a competitive relationship between swine and humans, Redding (1991) views pig avoidance from yet another perspective, arguing that the requirement of keeping pigs close to home puts them into direct competition with humans for basic food resources. In particular, Redding considers the species' penchant for raiding gardens and agricultural fields a significant threat to agricultural production. An increase in agricultural intensification, often associated with urban emergence, would exacerbate this competition and, he maintains, was a primary factor in the eventual abandonment of pig as an important meat resource in the Near East.

Finally, Diener and Robkin (1978) have traced the roots of pig avoidance in the region to elements of human political economy rather than swine physiology or behavior. They postulate that since pig rearing in the region was likely undertaken by small independent households, pigs did not fit well into centrally coordinated urban economies that sought to control the flow of commodities to dependent consumers. Moreover, the autonomy that small-scale pig production would afford independent households in such a controlled economy might be actively discouraged.

Of these four theories, only the first—which links pig avoidance to unfavorable environmental conditions—should function independently of the cultural conditions leading to greater social complexity in the region. The other three all specifically pertain to urban contexts. The development of specialist pastoral populations, the intensification of agriculture, and the development of centrally controlled complex economies are all part and parcel of the urban phenomenon in the ancient Near East. Therefore, the impact of these forces on pig keeping in the Near East should be particularly acute during periods of urban emergence and growth.

Case Examples of Pigs in Ancient Near Eastern Subsistence Economies

Tell Halif, Southern Levant

The first, and most detailed, case example of pig keeping in the ancient Near East discussed here is that of Tell Halif (Fig. 1). Halif is a small site located on the northern edge of the Negev desert in southern Israel, in a region that receives an average of 270 mm of rainfall per year. However, annual rainfall levels have been known to range from as much as 394 mm to as little as 141 mm. Thus Tell Halif is right around Grigson's rainfall cut-off point for pig raising, making it a prime candidate for examining the impact of shifting climatic conditions on pig utilization. Tell Halif is also an ideal location for examining the impact of urban development on subsistence. In addition to being located on the southernmost boundary for dry farming in the Levant, where settled agriculture is risky, the site is also situated on a major wadi system connecting Gaza with Hebron that has historically been a conduit for both economic relations and political brinkmanship. This combination of Halif's marginal environmental status and its placement on the political landscape probably accounts for major oscillations in the size and function of the site over its 5000 years of occupation (Table 1; Seger 1983, 1989, 1990).

The earliest evidence for occupation at Tell Halif dates to the Late Chalcolithic/Early Bronze I period, when a non-nucleated but still fairly sizable community was established on a low terrace northeast of the tell. Halif was apparently abandoned during the EB II period. The next major phase of occupation began during the early EB III, with the establishment of a nucleated, heavily fortified town. An ash layer over 3 m deep marks the destruction of this town sometime in the EB III A. After a short period of abandonment, a new town was built on the foundations of the razed settlement. Although lacking substantial fortifications, this town resembled the earlier settlement in both size and architectural sophisti-

Table 2. Distribution of species during different periods of occupation at Tell Halif (computed as percentages based on NISP of identifiable bones)

Period	Caprine %	Cattle %	Equid %	Pig %	Gazelle %	Birds %	Fish %	Total NISP
Iron Age	83	13	0.7	1	0.4	0.6	1	1418
Late Bronze	78	12	5	3	0.2	0.8	1	8243
Early Bronze III	92	5	0.3	0.3	1	0.9	0.4	5040
Early Bronze I	83	6	4	3	2	2	0.3	4291

cation. The three phases of occupation of this unwalled EB III B town are estimated to have lasted about 300 years. The site was once again abandoned in the EB IV and Middle Bronze periods during which time settlement in the vicinity shifted to Tell Biet Mirsim, 8 km to the north. Halif was reoccupied during the Late Bronze period and was the only settlement in the region. Based on the isolation of Halif, certain peculiarities in LB architectural features, and evidence of material culture ties with Egypt, Late Bronze Halif has been hypothesized to be a special function depot on a major trade route. Over an apparently continuous transitional period, the settlement at Halif evolved into a small fortified town during the Iron I (Philistine) and into the Iron II (Israelite) periods. Today Tell Halif is surrounded by Kibbutz Lahav.

Analysis of the close to 200,000 animal bones recovered during excavations at Tell Halif began under my direction in 1983. I was assisted in this work by Susan Arter, who now directs the analysis of animal remains recovered in excavations conducted at the site after 1992. A major focus of the research on the Halif fauna has been the response of animal exploitation to changes in the site's integration within local regional economies (Zeder 1990), with pigs, in particular, proving one of the more variable and interesting elements of the animal economy (Zeder 1996).

Table 2 presents the distribution of species in the various periods of occupation at the site (in the form of percentages based on NISP counts of identifiable bone). It is clear that in each period the species profile is heavily dominated by caprines (sheep and goat). However, there are some important shifts in species distributions over time. In particular, although caprines comprise 83% of the Early Bronze I assemblage, there is higher species diversity reflected in the remains from this diffuse settlement at the site than in any other period. The fauna from the Late Bronze trade outpost, which has the lowest proportion of caprines of any assemblage from the site (at 78%), also has a better representation of non-caprine domesticated and wild species than is found in the assemblages from either the Early Bronze III or Iron Age fortified towns. In contrast, the assemblage from the EB III town is especially noteworthy in its almost exclusive focus on caprines—92% of the assemblage of identifiable bones—and its overall low diversity of species. Similarly, in the Iron Age town, caprines and cattle together comprise 96% of the faunal assemblage.

Given the dominance of caprines in all periods of occupation at the site, it is useful to eliminate them from the picture and focus more closely on the distribution of other, non-caprine, species (Fig. 2). When this is done the most obvious trend in the data is the increase in the proportional representation of cattle at Tell Halif over time. Contributing only 6% of the assemblage from the EB I diffuse settlement at the site, cattle remain poorly represented in deposits from the EB III nucleated town, at only 5% of the identifiable bones. In contrast, their contribution more than doubles, to 12%, in the Late Bronze trade outpost, and then slightly increases to 13% of the assemblage from the Iron Age town site.

A somewhat more subtle pattern is the varying representation of equids and pigs in different occupational periods at Halif. Specifically, equids and pigs are relatively well represented in the EB I settlement remains and in the assemblage from the LB trade outpost, while their representation is comparatively depressed in the assemblages from the EB III and Iron Age town settlements at Halif. Two different phenomena can be tied to the peaks in equid representation in the EB I and LB periods. Remains in the EB I period belong primarily to a medium-sized equid, probably onager (*Equus hemionus*) or wild African ass (*Equus asinus*), that was hunted and eaten as a supplementary meat resource at the site.[1] In contrast, almost all the equid remains recovered from Late Bronze deposits at the site come from very small animals and are both metrically and morphologically ascribable to domestic ass (*Equus asinus*).

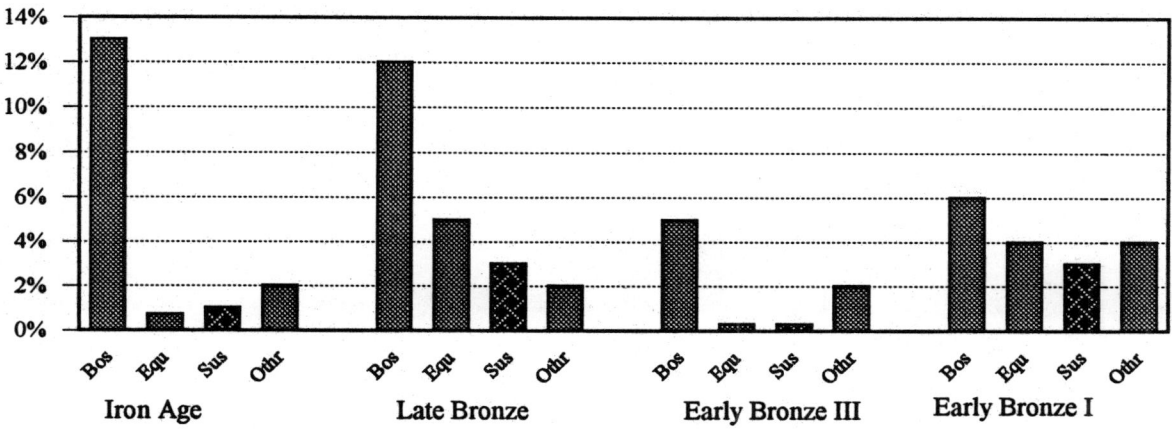

Fig. 2. Distribution of non-caprine species at Tell Halif; percentages based on NISP of identifiable bones.

Moreover, unlike the EB I equid remains (which were largely disarticulated, some with butchery scars), many of these equids were found as at least partially articulated skeletons and many were very young individuals. I have postulated elsewhere (Zeder 1990:27) that these equids were likely tied to Halif's Late Bronze Age function as a trading post and may, in fact, represent the local breeding of ass to be used in this trade network.

While a few of the pig remains from Tell Halif are quite large and likely represent wild boar, most of the remains represent small, young, domesticated swine. However, whether hunted or herded, swine were almost certainly consumed by Halif residents; therefore, the rise and fall in their representation within the assemblages from different occupational periods likely signals changes in the degree to which pigs were used as a food source at the site. In examining the distribution of pig bones over the various subphases of occupation (Fig. 3), it is clear that the peaks in the EB I and Late Bronze periods are not just an artifact of a couple of deposits that were particularly rich in pig bones within single subphases. Instead we see that the level of pig representation is either consistently high or consistently low in all subphases of a single phase of occupation. The paucity of pigs in the township occupations of Halif in the EB III and the Iron Age is also a general phenomenon in each subphase of these periods.

Thus the fluctuation of pig popularity at Tell Halif serves as a prime case example for evaluating the four different theories for pig avoidance in Near Eastern contexts discussed earlier. Each theory would attribute the fluctuations in the use of pigs at Tell Halif to different factors. Grigson's aridity theory would hold fluctuations in rainfall patterns in the northern Negev responsible for changes in the proportion of pigs utilized at Tell Halif.

Krader's nomadic theory would point to changes in the degree to which pastoral products were supplied by pastoral groups. Redding's agricultural intensification theory would attribute declines in pig exploitation to periods of regional agricultural intensification. And Diener and Robkin's urban economic theory would link the changing fortunes of pigs at the site to shifts in the degree to which Tell Halif was integrated into some kind of regional urban economy. It should be possible, then, to determine whether various factors linked to changes in the decline of pork consumption in the Near East (i.e., climatic fluctuations, nomadic pastoralism, agricultural intensification, or urban integration) vary with the degree of pig exploitation at Tell Halif, and so determine which of these different theories best accounts for the changing fortunes of pig at this site.

Grigson's aridity theory would require dramatic swings in climate between different phases in occupation. In particular, the Early Bronze I and Late Bronze, in which the proportion of pig remains was higher, should be relatively moister periods. In contrast, the EB III and Iron Age, which see significant declines in pig use, should be periods of greater aridity. Unfortunately, actual data on climatic conditions in the northern Negev during these periods are not available. However, settlement patterns in the region would not seem to support this scenario of climatic change. Periods with little pig use, which should be more arid and less able to support human settlement (EB III and Iron Age), are actually times of higher population density in this environmentally precarious region. During proposed wetter periods when pig utilization was higher, Halif was essentially the only settlement in the area (especially in the LB).

As to theories that link pig avoidance to the primacy of nomadic adaptations in the ancient Near East, there is

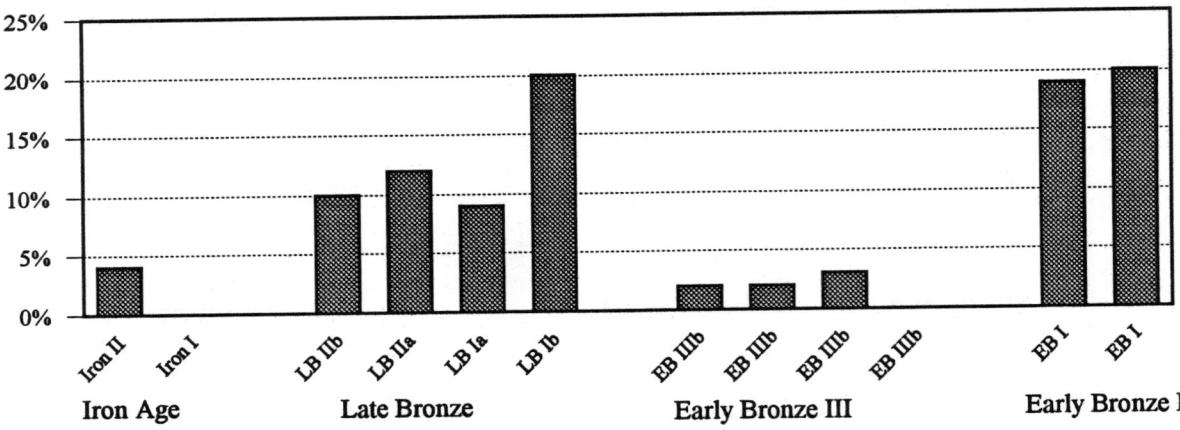

Fig. 3. Distribution of pig remains across different occupational phases at Tell Halif, based on normalized percents to compensate for different sample sizes from different strata.

some indication that nomads played a role in meat provisioning in the EB III period, a period in which pigs are particularly poorly represented. The increasingly tight focus on caprines, especially male caprines in the one-to-three-year age range, suggests a low degree of local involvement in animal production, which might in turn be linked to a higher degree of interaction with pastoral nomads (Zeder 1990:28–29, Zeder 1995a). Indeed, various researchers working in the region have pointed to pastoral nomads as important components of both the rise and collapse of Early Bronze urbanism in the Levant (Levy 1992; Dever 1980, 1989). During the Late Bronze period when pig utilization is once again relatively high, the broadening of the diversity of species and the presence of a large number of very young, less than six months old, sheep and goats argues for a greater degree of local involvement in herd management (Zeder 1990:27). However, there is also strong evidence for a local focus in animal production during the Iron Age, when pig use is once again low. Obviously something more complex than the role of pastoral nomads in provisioning the site is at play here.

Interestingly, lows in pig popularity do in fact coincide with times of higher population density in the region, when an intensification of local agricultural production is likely. This pattern seemingly supports Redding's agricultural intensification theory. Redding's theory also predicts that while the importance of pigs should decrease with intensification of agricultural economy, there should be a corollary increase in the importance of cattle. Unlike pigs, which when not confined pose a significant threat to fields, cattle are used for both plowing and threshing grain and are thus important contributing elements in the intensification of agricultural production. Moreover, increased use of cattle dairy products would offset the nutritional loss incurred by decreasing reliance on pigs. When the ratios of cattle bones to pig bones in different periods of occupation at the site are compared (Table 3), we see that, as Redding would predict, there is a strong increase in the number of cattle bones relative to pig bones in periods of presumed intensification of agricultural practices, and a decrease during periods when agriculture was likely less intensified.

However, there are aspects of the data from Tell Halif that do not follow other predictions generated by Redding's model. Since cattle and sheep prefer the same kinds of pasture plants, Redding argues that an increase in the importance of cattle will put cattle and sheep into direct competition for pasture resources. Goats, on the other hand, can subsist on a different range of pasture plants which are eschewed by their more discriminating bovine cousins, and can thus be effectively grazed on the same pastures as cattle. Therefore, Redding predicts that with agricultural intensification there should not only be an increase in cattle and a decrease in the number of pigs utilized, but also an increase in the representation of goats relative to sheep. Data from Tell Halif do not follow this predicted pattern (Table 3). In fact, sheep are especially well represented relative to goats during the two periods when the greatest degree of agricultural intensification is expected (the EB III and Iron Age), and they are most poorly represented during periods that might be expected to have seen a fairly low level of agricultural intensification in the region (EB I and LB).

It is also instructive to look at the ratio of caprine to cattle bones in various phases of occupation at Halif (Table 3), an examination which brings into sharper focus patterns apparent in the tabular and graphic presentation of relative data in Table 2 and Figure 2. Rather than a correlation between increased cattle utilization and periods of likely agricultural intensification at the site (EB III and Iron Age), and the reverse during periods in which Halif operated relatively independently (EB I and LB), there is no apparent correlation between cattle representation and probable periods of more intensified agricultural production. Instead, the proportion of cattle relative to caprines is quite low during both the EB I and EB III periods, and much higher in the later LB and Iron Age phases of occupation.

Table 3. Ratios of various species at Tell Halif (computed from NISP of identifiable bones)

Period	Cattle:Pig	Sheep:Goat	Caprine:Cattle
Iron Age	13.5	1.6	6.2
Late Bronze	3.9	1.0	6.7
Early Bronze III	16.6	1.6	18.7
Early Bronze I	2.0	1.4	14.4

One of the problems with Redding's model is that it does not take into account alternative pasturing regimes for livestock. Competition between livestock for pasturage could be avoided by pasturing these animals in different areas, by providing supplemental fodder, or, especially in the case of pigs, penning animals. And while large-scale sty management of pigs in arid environments requires more sophisticated technological approaches to livestock management than were available to ancient Near Eastern herdsmen, pigs could have been quite profitably managed on a small-scale, sty-bound basis that was managed on the individual household level. This strategy would effectively eliminate the threat that pigs pose to agricultural fields and gardens, while allowing for a fairly high collective level of pig production among independent households. Redding's theory may also fall short in its concentration on the process of agricultural intensification rather than on the overarching economic processes that motivate this intensification, along with other kinds of economic specialization, particularly the development of a regional, possibly centrally coordinated, urban-based economy.

Indeed, it is the degree of integration within a well developed, possibly centrally coordinated economy that seems to vary most consistently with the fortunes of pigs at the site, suggesting that the Diener and Robkin political economy model best fits the Tell Halif case study example. During periods of higher pig use, Halif was a relatively isolated, self-sufficient site. The diffuse EB I settlement at Halif predates the establishment of a fairly tightly knit late EB I/EB II regional economy focused on centers like Tell Arad to the south. Here pigs were part of a broad-based diet that utilized both domestic animals and a relatively wide array of wild species. During the Late Bronze period, when Halif was an isolated trade outpost, pigs would seem to have been an easily produced supplementary food resource for Halif residents involved in conducting the business of this special function settlement.

On the other hand, pigs are less common or absent altogether in periods of greater integration within a regional urban economy. Pigs are entirely absent in the earliest EB III levels at Halif—a period argued to represent the apogee of urban integration in southern Palestine. They reappear, but only in very small numbers, in later phases of the EB III during periods of regional urban disintegration (Fig. 3; Zeder 1995b, 1996). In the Iron Age, when the fortified town at Halif is well integrated into the Judean royal system, pigs are again poorly represented. Although this last decline in pig popularity coincides with the Israelite period and the possible levying of dietary laws that prohibit consumption of swine flesh, the earlier fluctuations in pig use suggest that such dietary proscriptions alone cannot account for this pattern.

The case example of Kibbutz Lahav located next to Tell Halif demonstrates the ongoing role that pigs play in the political economy of the current-day Israeli state. After experimenting with the management of cattle, sheep, and goats, residents of Kibbutz Lahav have settled on pig as their primary animal crop. With the help of modern technology, they are able to raise pigs on a relatively large scale. Kept in segregated pens in air conditioned barns with ready access to water, shade, and wallow, pigs have proven a highly reliable and efficient meat source for Kibbutz kitchens. Most aspects of the local economy at Kibbutz Lahav are tied to the larger Israeli economy. A factory on the kibbutz makes plastic goods that are marketed elsewhere. The turkey hatchlings brought to the kibbutz to be fattened, are shipped to off-site wholesalers and sold by non-kibbutz retailers. In contrast, pig production and marketing at Kibbutz Lahav is a highly localized enterprise from start to finish. Pigs are bred, reared, fattened, butchered, and eaten by kibbutz residents. Marketing of Lahav pigs to non-kibbutz residents takes place on or adjacent to kibbutz grounds, and is conducted by kibbutz members. This unorthodox community persists in this economic enterprise in direct defiance of Jewish dietary law and despite threatened sanctions from more conservative elements of the Israeli state. Kibbutz Lahav's investment in pigs thus provides a dramatic present-day example of the use of pigs in this region to maintain a level of economic autonomy in an otherwise highly controlled economy.

I would argue, then, that the fluctuating presence of pigs in the diet of Tell Halif residents cannot be attributed to variable environmental conditions. Nor can these dietary changes be linked primarily either to cyclical variations in the role of nomads in the economy, or to the changes in the degree of agricultural intensification in the region. Instead, the one factor that best explains the shifts in pig utilization at the site is the degree to which the site is integrated into a broader regional economy. These results are echoed by those of Hesse and Wapnish (Hesse 1990, 1994; Hesse and Wapnish 1987), who found that in the Middle to the Late Bronze Age pig utilization in the Levant was inversely related both to the growth of a large centralized olive processing industry, and to the degree of political and economic domination by external powers. Taken together, these studies suggest that the connection seen between decreased utilization of pigs and the presence of an urban economy at Halif may be a more widespread phenomenon in the urban ancient Near East.

Mesopotamia

Given the patterns seen in the Levant, it is instructive then to broaden our focus on pig utilization in Near Eastern urban contexts and to review data on pigs in early urban systems in greater Mesopotamia. For the past 10 years I have been working with the faunal remains from a number of sites in the Khabur Basin of northeastern Syria (Zeder 1994a, 1995b). The sites date from about 7000 B.C. to about 1500 B.C. and span the period from the first introduction of domesticated crops and livestock into the region up through the rise of the first state-level societies in northern Mesopotamia (Fig. 1). The goal of this research is to examine both the environmental and cultural impact of the introduction and intensification of herding and farming in a geographically discrete but environmentally varied region over a long temporal span. Of particular interest here is the process leading up to the emergence of urban economy in the region, and how the changing nature of pig utilization in the region relates to this process.

From the mid-eighth through the fourth millennium B.C., the Khabur Basin experienced a gradual growth of population, especially in the better watered northern steppe where a network of agricultural villages and towns was established. Population levels in the more arid southern steppe remained quite low, with only a few isolated sites occupied at any one time, these usually situated on the floodplain of the Khabur river. Archaeobiological analysis of plants and animals from sites covering this time period point to the existence of a number of highly localized subsistence strategies. In the northern, more densely settled steppe there was an emphasis on domesticated crops and livestock—despite strong indications that herds of gazelle, onager, and aurochs may have remained quite plentiful there. In contrast, the isolated sites in the more arid southern steppe zone utilized different blends of wild and domestic resources. Wild animals were often the predominant source of meat, with each site having its own distinctive mixture of wild and domestic resources.

There was a fairly substantial increase in settlement in the Khabur in the early third millennium B.C., a process that culminated in the establishment of large urban centers supported by highly structured settlement systems in the northern steppe by the mid-third millennium (Stein and Wattenmaker 1990). Archaeobiological evidence from later third and second millennium contexts points to a concomitant shift from highly localized economies to a region-wide subsistence economy focusing primarily on specific cereal crops and domesticated livestock (Zeder 1995b:29). The mechanism by which these localized strategies were replaced is a central focus of ongoing research in the region. Were localized subsistence practices suddenly replaced with a more specialized, regionally integrated subsistence economy, or did this later third millennium economy evolve more slowly out of long-standing local traditions? When did this shift occur? And what can this transformation period tell us of the relationship between pig keeping and urban emergence in the ancient Near East?

Faunal remains from four small early to mid-third millennium sites in the southern portion of the region provide a rare opportunity to monitor the process of increasing specialization of subsistence economy in the Khabur Basin (Zeder 1998). Third millennium occupations at all four sites—Tell Ziyadeh, Tell Raqa'i, Tell Atij, and Tell Gudeda (see Fig. 1)—were established as part of a remarkable proliferation of similar small settlements along the banks of the southern Khabur River (e.g., Kerma, Mullah Mutar, Bderi). Excavations at Tell Atij and Tell Raqa'i have documented four roughly contemporary levels of third millennium occupation that span the period from the early to the mid-third millennium B.C. (Fortin 1993:106). Recent excavations at Tell Ziyadeh have encountered a single phase of third millennium occupation that likely overlaps with the early levels of occupation at Tell Raqa'i and Tell Atij (F. Hole, pers. comm. July 1997). Tell Gudeda, located across the river from Tell Atij, seems to have been occupied somewhat later, overlapping only slightly, if at all, with the final stages of occupation at Atij (Fortin, Routledge, and Routledge 1994; Fortin, pers. comm. August 1995). Like other third millennium settlements in the region, all these sites are characterized by the presence of large granaries or other special function structures that suggest some role in broader regional economic interactions. Some researchers have suggested that these riverside settlements stored grain that was being transshipped to large urban centers to the south, most notably the site of Mari (Schwartz 1993, 1994; Fortin 1991). Others have posited that the storage facilities at these sites held grain both to feed an increased number of pastoral peoples in the region and to be used as fodder for their flocks (Hole 1991). McCorriston has, in turn, linked these pastoralists to the burgeoning textile industries based in contemporary urban centers (1995, 1997).

There are striking similarities in the assemblages from these sites, and together they provide unique insight into the transformation of subsistence economies in the southern steppe over this crucial transitional period leading up to urban emergence in the Khabur Basin. Basal levels of occupation at Raqa'i, Atij, and the assemblage from Ziyadeh, for example, echo earlier patterns of animal utilization in the region (Fig. 4). At each site there is a distinctive mixture of wild and domestic resources, with particularly strong emphases on wild game at basal levels at Atij and at Ziyadeh. At Ziyadeh, however, gazelle was the predominant wild animal utilized, while onager seems to have been a particularly important resource at Tell Atij.

Of special interest here is the uniformly heavy emphasis on pig seen at all three sites. Pig remains constitute 25% of the third millennium deposits at Ziyadeh and about 20% of the assemblages from early levels at both Raqa'i and Atij. Successive levels of occupation at Raqa'i and Atij indicate a gradual replacement of traditional broad-based strategies by an increasingly exclusive focus on domesticated caprines. At both sites the contribution of pigs to the faunal assemblages drops from about 20% in early levels to only 1–2% in the final levels of occupation. At Atij there is also a sharp reduction in the importance of onager, from nearly 30% of the assemblage in basal layers to only 10% in that from the final occupation layers. Although onager never seems to have played as important a role at Tell Raqa'i, even there a reduction in onager over time is evident, from a high of 9% of the assemblage in the basal levels to only 2% in the final occupation level. In contrast, there is a steady increase in the proportional representation of caprines at both sites: at Tell Atij, from 33% of the assemblage in basal levels to nearly 70% in the final levels of occupation, and at Raqa'i, from 65% in Level 5 to 86% of the Level 2 assemblage. The faunal remains from Tell Gudeda provide a capstone to this regional shift in animal exploitation. Caprines are overwhelmingly the predominant animals represented at the site, with pig, onager, and gazelle only very minor components of the subsistence economy.

The reduction in the utilization of wild animals at Atij and Raqa'i might to some extent reflect the beginning of the process of human-induced environmental deterioration, which culminated in the virtual disappearance of these once dense herds from the region. Modification of the region by expanded human populations, however, seems to have affected different wild species in different ways. For example, while the representation of onager falls dramatically over time, the pro-

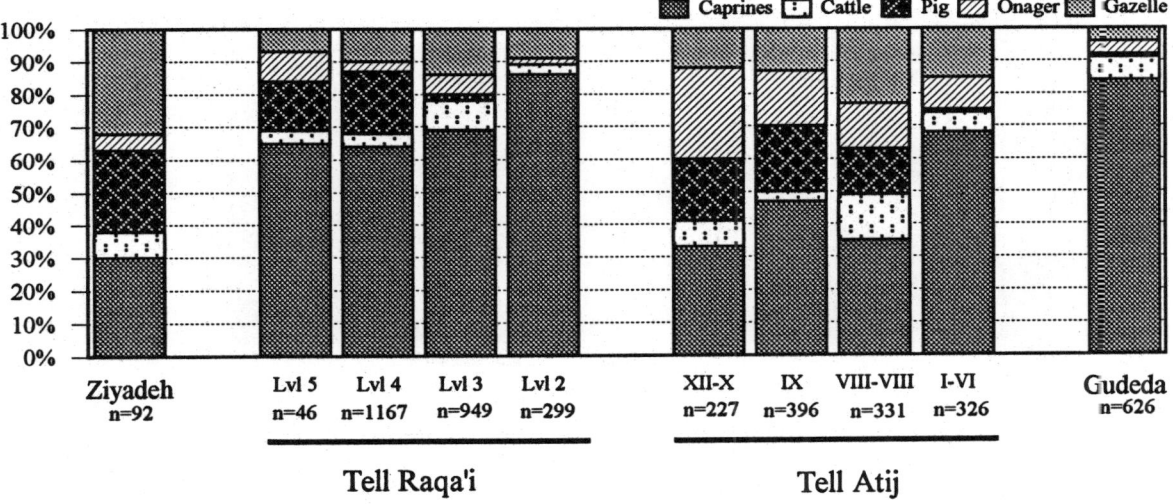

Fig. 4. Distribution of species through time at Tell Ziyadeh, Tell Atij, and Tell Raqa'i.

portion of gazelle utilized at both sites is relatively constant. Clearly there was not a wholesale destruction of the steppe during the early to mid-third millennium B.C. It is possible that onager populations were more sensitive to whatever disruptions the increased human populations presented to the ecology of the southern steppe. It is also possible, however, that the abandonment of onager hunting was the result of an economic restructuring of subsistence practices in this region.

Certainly changes in utilization of domesticated animals at these sites would argue that factors beyond the impact of expanded populations on fragile environments were operating here, and would suggest that economic forces played a more significant role in the evolution of third millennium subsistence in the southern steppe. Specifically, the decline in pig utilization, the low levels of cattle in these assemblages, and the increasingly exclusive focus on caprines point to the transformation of localized, broad-based subsistence strategies that had operated in the area for several millennia into a more specialized pastoral economy. By the mid-third millennium, when large sites dominated the northern steppe, it would appear that the southern steppe had become incorporated into a region-wide economy in which areas of lower agricultural potential were dedicated to the support of expanded herds of sheep and goats. These caprine herds, in turn, supplied a variety of pastoral products to residents of the cities, towns, and villages located in the more prime agricultural areas of the north.

Pig remains also provide insight into some of the dynamics that shaped social and economic relations in northern urban contexts in the Khabur Basin. While northern sites of the mid- to late third millennium shared the south's almost exclusive focus on domesticated animals, residents of northern sites seem to have utilized a broader suite of domesticated livestock. In particular, pork consumption seems to have been more common here than in contemporary southern contexts. There is also good evidence, however, of socio-economic variation in the use of pig in different urban contexts in the north. Pigs are surprisingly well represented at an elite/sacred area at Tell Leilan, one of the major mid-third millennium urban center in the Khabur. Ongoing analysis of the Leilan remains that I am directing indicates that pigs constitute 32% of the domestic species from third millennium deposits on the acropolis of this large tell (Fig. 5; Stackhouse 1998). However, pork consumption seems to have been even more common in more humble residential sectors of the site. Pig bones make up nearly 50% of the remains of domesticates recovered from the lower town area of shops and small houses. It would seem then that while pigs were an important supplementary resource to the elite diet at Leilan, they were the mainstay of the urban diet among Leilan shopkeepers, merchants, and urban poor.

This pattern is reminiscent of the status-related variation in pork consumption noted by Mudar at third millennium levels at ancient Lagash, modern al-Hibba, in southern Mesopotamia (Fig. 5; Mudar 1982). In her analysis of the faunal remains from al-Hibba, Mudar found that while pigs made up only 8% of the faunal remains from temple contexts, they comprised almost 20% of the remains recovered in residential sectors of the site. This emphasis on pigs in non-elite contexts seems to

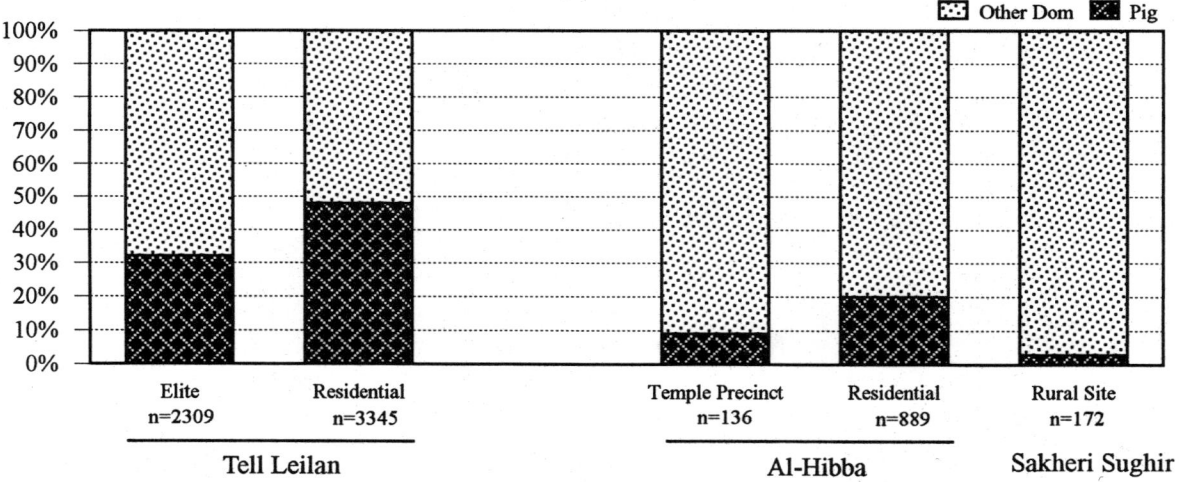

Fig. 5. Comparison of pigs to other domestic animals in different urban and rural contexts in northern and southern Mesopotamia (Bökönyi and Flannery 1969; Mudar 1982).

be a particularly urban phenomenon, judging by the faunal remains recovered from the nearby rural community of Sakheri Sughir, where pig bones made up only 3% of the assemblage of analyzed bones (Bökönyi and Flannery 1969). All this evidence bolsters the conclusion that pigs were not a part of the large, specialized animal economy of early Mesopotamian city-states, and that neither the elites provisioned by this economy nor the rural producers who supplied these resources depended on pigs in a major way. Instead, utilization of pigs was primarily confined to poorer urban contexts, where these animals were likely raised by local households as a means of supplementing food resources received through other channels.

Additional evidence for pig production being a local level, non-regulated part of the pastoral economy in southern Mesopotamia is found in the general absence of references to pigs or pig keeping in economic texts from the region (Postgate 1992:166). One particularly striking example of this can be found in the textual data from the late third/early second millennium Ur III animal distribution center at Puzrish Dagan (modern Tell Drehem). Economic texts recovered from this site record, in meticulous detail, the numbers, species, and even breeds of animals that were channeled into this center, transferred through at least two administrative levels, and distributed to a variety of very specific recipients—soldiers, temples, prominent individuals, kennels (Jones and Snyder 1961; Kang 1972). In an earlier study I tabulated records covering essentially the entire span of time that the center was in operation (Zeder 1994b). Using data from these texts I could trace the flow of more than 10,000 animals into, through, and out of this center. Distinctive profiles in the age, sex, and species of animals were identified. Moreover, different groups of recipients received distinctly different types of animals from Drehem. Young lambs and kids were particularly well represented in disbursements made to temples, cattle and older ewes were directed to soldiers, and donkeys went exclusively to the kennels.

And yet for all their specificity, these texts are perhaps most notable for what they do not contain. First, the list of recipients of Drehem animals is limited to people likely to have served directly at the pleasure of the state. There is no mention of disbursements to farmers, herders, or even to craftsmen, who together must have comprised the bulk of the Ur III economy. Second, the list of animals distributed is limited primarily to sheep and goats, a few cattle, asses, and some wild species, primarily gazelle. There is no mention of fish, which we know to have been an important, and at least partially regulated, food resource in southern Mesopotamia (Salononen 1970). Nor is there any mention of pigs. None of the hundreds of transactions I used in this earlier study referred to pigs or pig products. Yet we know from the faunal remains recovered from the urban sites mentioned above, as well as from other southern Mesopotamian sites (e.g., Mudar 1982; Bökönyi 1978:189), that pigs were consumed, and sometimes in quantity, in certain urban settings. Thus the absence of pigs from the detailed accounts of the bureaucrats responsible for the collection and disbursement of animals from this state-controlled animal clearing house helps to make the case for pigs being an unregulated animal resource in urban contexts in the Near East.

Conclusions

In this paper I have tried to isolate the most significant factors that contributed to the decline and eventual rejection of pig as a primary animal resource in the ancient Near East. Certainly it is the unsuitability of pigs for a role in large-scale redistributive economies which seems the most important factor in this process. In reality, however, it is a larger constellation of environmental, physiological, and behavioral constraints on pig keeping in arid environments that makes the pig such an unlikely candidate for a starring role in this kind of economy in the Near East. This combination of factors dictates the conditions of pig keeping in these environments, and makes pigs ill-suited to long-distance transhumance, to free-range pasturing, and to large-scale centrally controlled production. Owing to this combination of water and food requirements, heat tolerance, and behavior, small-scale sty-based management by individual urban households was the optimal and perhaps only effective strategy available for taking advantage of this animal's remarkable meat-producing abilities in ancient Near East urban societies. At the same time, pigs provided a means of buffering some of the uncertainties of provisioning garnered through either market or state-controlled redistributive mechanisms. In fact, the degree to which small-scale pig rearing provided a buffer against the vicissitudes of state-level provisioning, and offered individual households some measure of economic autonomy, may well have made pigs a potentially seditious force in urban economies—one which those interested in controlling the flow of marketable commodities might well have wanted to discourage.

By extension, this constellation of factors can also be linked to the eventual proscription of pork by the followers of two of the major religions in the region. Proscriptions serve an important role as markers of social boundaries, and as a means of consolidating social solidarity and corporate identity. They are a visible and enforceable means of demonstrating membership within a group, and of drawing distinctions between group members and outsiders. The choice of specific food items as candidates for proscription is no doubt shaped by a variety of factors, many of which may have little apparent functionalist rationale. However, there is little reason to believe that these choices are made in a functional vacuum, without any reference to the environmental and economic parameters that shape the productivity or desirability of a food resource in a particular circumstance. In the case of proscriptions concerning the consumption of pork, the fact that pig rearing may have been primarily the independent purview of urban, non-elite households would seem to have made pigs a more likely candidate for proscription than an animal tightly linked to large-scale urban provisioning systems. Moreover, since pigs offer a limited range of resources than other major livestock species in the region, their classification as unclean would have a more narrowly focused economic impact. In addition, the identification of pigs with sedentary village or lower class urban dwellers might have contributed to the dietary exclusion of pork by Jews and Muslims, whose social identity relies heavily on their nomadic heritage. Here the physiological and morphological differences that distinguish pigs from animals herded by nomads (i.e., the fact that pigs have cloven hoofs like sheep and goats but do not chew their cuds) would take on a special significance.

It is clear that the inclusion, or exclusion, of pork from the diet in the ancient Near East carries information that goes far beyond the realms of dietary preference or environmental parameters. Indeed, the level of pork consumption in Near Eastern urban diets might well be taken as a measure of the effectiveness of a regional economy in coordinating the flow of food staples, and the degree to which indirect provisioning systems meet urban food needs. Likewise, the presence of pigs in certain urban settings may be a marker of an effort to maintain a level of economic autonomy within an otherwise highly controlled economy. The possibility that household-based swine management in urban contexts was the domain of women has interesting, as yet unexplored, implications for those interested in the role of gender in emergent complexity. The religious proscription of pork also has important bearing on our understanding of the forces that shape ideological belief systems, and on the role of ideology in codifying social and economic relations in early urban contexts. Thus, rather than a dietary dead end, pigs should instead be viewed as one of the more interesting and illuminating aspects of subsistence economy in the ancient Near East. In fact, pigs are perhaps best viewed as a multi-dimensional index of social and economic cohesion that allows us to examine some of the basic forces which shape economy, polity, and ideology not only in the ancient Near East, but (as amply demonstrated by the wide-ranging contributions in this volume) in other areas of the world as well.

Notes

1. There are a few exceedingly large equids from Late Chalcolithic/EB I levels at the site that may be horse. These bones have distinctive cabelline morphological characteristics that resemble, in size, specimens reported by Grigson (1993), Davis (1976), and Lernau (1978) from the southern Levant. The possible presence of horse in this region at this early date has important zoogeographic and cultural implications that will be explored in a later publication.

References

Bökönyi, S. 1978. The Animal Remains from the 1970–1972 Excavation Seasons at Tell el-Der: A Preliminary Report. In *Tell el-Der II: Progress Reports (First Series)*, ed. L. De Meyer, pp. 185–189. Editions Peeters, Leuven.

Bökönyi, S, and K.V. Flannery. 1969. Report on the Animal Remains from Sukheri Sughir. In *The Administration of Rural Production in an Early Mesopotamian Town*, H.T. Wright, pp.143–149. Museum of Anthropology, University of Michigan, Anthropological Papers, No. 38. Ann Arbor.

Davis, S. 1976. Mammal Bones from the Early Bronze Age City of Arad, Northern Negev, Israel: Some Implications Concerning Human Exploitation. *Journal of Archaeological Sciences* 3:153–164.

Deever, W.G. 1980. New Vistas on the EB IV ("MB I") Horizon in Syria-Palestine. *Bulletin of the Schools of Oriental Research* 237:35–64.

——— 1989. The Collapse of the Urban Early Bronze Age in Palestine—Toward a Systemic Analysis. In *L'urbanisation de la Palestine à l'âge du bronze ancien: Bilan et perspectives de recherches actuelles*, ed. P. De Miroschedji, pp. 225–246. BAR International Series No 527. Oxford.

Diener, P., and E.E. Robkin. 1978. Ecology and Evolution and the Search for Cultural Origins: The Question of Islamic Pig Prohibition. *Current Anthropology* 19:493–540.

Douglas, M. 1966. *Purity and Danger: An Analysis of Concepts of Pollution and Taboo.* Prager, New York.

Flannery, K. 1983. Early Pig Domestication in the Fertile Crescent: A Retrospective Look. In *The Hilly Flanks: Essays on the Pre-History of Southwest Asia*, ed. T.C. Young, P.E.L. Smith, and P. Morteson, pp. 163–188. Studies in Ancient Oriental Civilization 36, Oriental Institute. University of Chicago, Chicago.

Fortin, M. 1991. Récentes recherches archéologiques dans le Moyenne Vallée du Khabour (Syrie). *Bulletin of the Canadian Society for Mesopotamian Studies* 21:5–16.

——— 1993. Résultats de la 4ème campagne de fouilles à Tell 'Atij et de la 3ème à Tell Gudeda, Syrie. *Echos du monde classique/Classical Views* 37:97–121.

Fortin, M., B. Routledge, and C. Routledge. 1994. Canadian Excavations at Tell Gudeda (Syria) 1992–1993. *Bulletin of the Canadian Society for Mesopotamian Studies* 27:51–64.

Grigson, C. 1987. Pastoralism, Pig-Keeping and Other Aspects of the Chalcolithic in the Northern Negev. In *Shiqmim I*, ed. T.E. Levy, pp. 219–241. British Archaeological Reports, International Series 356. Oxford.

——— 1993. The Earliest Domestic Horses in the Levant? New Finds from the Fourth Millennium of the Negev. *Journal of Archaeological Science* 20:645–655.

Harris, M. 1974. *Cows, Pigs, Wars, and Witches: The Riddles of Culture*. Random House, New York.

——— 1985. *Good to Eat: Riddles of Food and Culture*. Simon and Schuster, New York.

Hatt, R.T. 1959. *The Mammals of Iraq*. Miscellaneous Publications, Museum of Zoology, 106. University of Michigan, Ann Arbor.

Hesse, B. 1990. Pig Lovers and Pig Haters: Patterns of Palestinian Pork Production. *Journal of Ethnobiology* 10:195–225.

——— 1994. Husbandry, Dietary Taboos, and the Bones of the Ancient Near East: Zooarchaeology in the Post-Processual World. In *Methods in the Mediterranean*, ed. D. B. Small, pp. 197–232. E.J. Brill, Leiden.

Hesse, B., and P. Wapnish. 1987. Pig Avoidance in the Iron Age. Paper presented at the Annual Meeting of the American Schools of Oriental Research, Boston, December.

Hole, F.A. 1991. Middle Khabur Settlement and Agriculture in the Ninevite 5 Period. *Bulletin of the Canadian Society for Mesopotamian Studies* 21:17–30.

Jones, T.B., and J.W. Snyder. 1961. *Sumerian Economic Texts from the Third Ur Dynasty: A Catalogue and Discussion of Documents from Various Collections*. University of Minnesota Press, Minneapolis.

Kang, S.T. 1972. *Sumerian Economic Texts from the Drehem Archive: Sumerian and Akadian Texts in the Collection of the World Heritage Museum of the University of Illinois*, Vol. 1. University of Illinois Press, Urbana.

Krader, L. 1955. Ecology of Central Asian Pastoralism. *Southwestern Journal of Anthropology* 11:301–326.

Lernau, H. 1978. Faunal Remains, Strata III–I. In *Early Arad*, ed. R. Amiran, pp. 83–113. Israel Exploration Society, Jerusalem.

Levy, T.E. 1992. Transhumance, Subsistence, and Social Evolution in the Northern Negev Desert. In *Pastoralism in the Levant: Archaeological Materials in Anthropological Perspectives*, ed. O. Bar-Yosef and A. Khazanov, pp. 83–92. Monographs in World Archaeology No. 10. Prehistory Press, Madison, WI.

McCorriston, J. 1995. Preliminary Archaeobotanical Analysis in the Middle Habur Valley, Syria, and Studies of Socioeconomic Change in the Early Third Millennium BC. *Bulletin of the Canadian Society for Mesopotamian Studies* 29:33–46.

——— 1997. The Fiber Revolution: Textile Extensification, Alienation, and Social Stratification in Ancient Mesopotamia. *Current Anthropology* 38:517–549.

Mount, L.E. 1968. *The Climatic Physiology of the Pig*. Williams and Watkins, Baltimore.

Mudar, K. 1982. Early Dynastic III Animal Utilization in Lagash: A Report on the Fauna from Tell Al-Hibba. *Journal of Near Eastern Studies* 4:23–34.

Perry, J.H. 1985. Early European Penetration of Eastern North America. In *Alabama and the Borderlands: From Prehistory to Statehood*, ed. R.R. Badger and L.A. Clayton, pp. 83–96. The University of Alabama Press, Tuscaloosa.

Postgate, J.N. 1992. *Early Mesopotamia: Society and Economy at the Dawn of History*. Routledge, London.

Redding, R. 1991. The Role of Pig in the Subsistence System of Ancient Egypt: A Parable on the Potential of Faunal Data. In *Animal Use and Culture Change*, ed. P.J. Crabtree and K. Ryan, pp. 20–30. MASCA Research Papers in Science and Archaeology, Supplement 8. University of Pennsylvania Museum, Philadelphia.

Salononen, A. 1970. *Die Fischerei im Alten Mesopotamia nach Sumerisch-Akkadischen Quellen*. Annales Academiae Scientiarum Fennicae, series B, 149. Helsinki.

Schwartz, G.M. 1993. Rural Archaeology in Early Urban Mesopotamia. *National Geographic Research and Exploration* 9:120–131.

——— 1994. Rural Economic Specialization and Early Urbanization in the Khabur Valley, Syria. In *Archaeological Views from the Countryside: Village Communities in Early Complex Societies*, ed. G.M. Schwartz and S.E. Falconer, pp. 19–36. Smithsonian Institution Press, Washington, DC.

Seger, J. 1983. Investigations at Tell Halif, Israel, 1976–1980. *Bulletin of the American Schools of Oriental Research* 252:1–23.

——— 1989. Some Provisional Correlations in EB III Stratigraphy in Southern Palestine. In *L'urbinisation de la Palestine à l'âge du bronze ancien*, ed. P. de Miroschedji, pp. 117–135. British Archaeological Reports, International Series 527(ii). Oxford.

——— 1990. The Bronze Age Settlements at Tell Halif: Phase II Excavations, 1983–1987. *Bulletin of the American Schools of Oriental Research*, Supplement 26:1–32.

Stackhouse, S. 1998. The Effects of Socio-Economic Status on Dietary Patterns in the Emergent Urban Society of Tell Leilan. Unpublished manuscript prepared for the National Museum of Natural History Research Training Program, August 1998.

Stein, G., and P. Wattenmaker. 1990. The 1987 Tell Leilan Regional Survey: Preliminary Report. In *Economy and Settlement in the Near East: Analysis of Ancient Sites and Materials*, ed. N. Miller, pp. 8–18. MASCA Research Papers in Science and Archaeology, Supplement to Vol. 7. University of Pennsylvania Museum, Philadelphia.

Van Loon, D. 1978. *Small-scale Pig Raising*. Storey Communications, Pownal, VT.

Weiss, H. 1986. The Origins of Tell Leilan and the Conquest of Space in Third Millennium Mesopotamia. In *The Origins of Cities in Dry-Farming Syria and Mesopotamia in the Third Millennium BC.*, ed H. Weiss, pp. 71–108. Four Quarters Publishing, Guilford, CT.

Zeder, M.A. 1990. Animal Exploitation at Tell Halif. *Bulletin of the American Schools of Oriental Research*, Supplement No. 26:24–30.

——— 1994a. After the Revolution: Post-Neolithic Subsistence Strategies in Northern Mesopotamia. *American Anthropologist* 96(1): 97–126.

——— 1994b. Of Kings and Shepherds: Specialized Animal Economy in Ur III Mesopotamia. In *Chiefdoms and Early States in the Near East: The Organizational Dynamics of Complexity*, ed. Gil Stein and Mitchell Rothman, pp. 175–191. Prehistory Press, Madison, WI.

——— 1995a. The Desert and the Sown: Pastoral/Urban Interaction in the Early Bronze Age Southern Levant. Paper delivered at the Annual Meetings of the Society of Ethnobiology, Phoenix, AZ, March.

——— 1995b. The Archaeobiology of the Khabur Basin. *Bulletin of the Canadian Society for Mesopotamian Studies* 29:21–32.

——— 1996. The Role of Pigs in Near Eastern Subsistence from the Vantage Point of the Southern Levant. In *Retrieving the Past: Essays on Archaeological Research and Methodology in Honor of Gus Van Beek*, ed. J.D. Seger, pp. 297–312. Eisenbrauns/Cobb Institute of Archaeology, Winona Lake, IN.

——— 1998. Environment, Economy, and Subsistence on the Threshold of Urban Emergence in Northern Mesopotamia. In *Espace Naturel, Espace Habité en Syrie Nord (10ᵉ–2ᵉ millénaire av. J.-C.)*, ed. M. Fortin and O. Aurenche, pp. 55–67. Bulletin of the Canadian Society for Mesopotamian Studies 33 and Travaux de la Maison de l'Orient 28. The Canadian Society of Mesopotamian Studies, Québec; La Maison de l'Orient Méditerranéen, Lyon.

Zeuner, F. 1963. *A History of Domestic Animals*. Hutchinson & Co., London.

PIG USE AND ABUSE IN THE ANCIENT LEVANT: ETHNORELIGIOUS BOUNDARY-BUILDING WITH SWINE

Brian Hesse and Paula Wapnish

Department of Anthropology, University of Alabama at Birmingham, Birmingham, AL 35294-3350

Introduction

It takes a bit of *chutzpah* for a pair of zooarchaeologists to embark on a discussion of the ethno-religious significance of pigs in the ancient Near East and, in particular, to examine the meanings and behaviors associated with the animal in the Levantine heartland where the biblical texts that command a people to avoid swine were composed. After all, not only have such influential voices in modern anthropology as those of Mary Douglas (1966) and Marvin Harris (1985) offered thoughts on the issue, but from the recent major syntheses of Jacob Milgrom (1991) and Walter Houston (1993), the path of textual and theological treatments of the food laws leads back to Maimonides and earlier to the rabbis who produced the Talmud and the Mishnah.[1] Given this mountain of literature, what more could possibly be left to be said?

To answer our own question, we intend to sidestep the contentious search for the one "true" rationale that lies behind the origin of pig disdain as expressed in the Bible. This debate has proposed, among other things, that the Israelite avoidance of swine is a moral teaching designed to promote moderation, a hygienic rule contributing to health, a reaction to swine behavior and appearance, a way to avoid a serious disease, a logical precipitate of the folk classification system, a response to diminished forest cover, a link to Egyptian theological concepts, or a tactic to insure a successful migration to the Promised Land. We will explore instead how pig remains found (or not found) on archaeological sites might be interpreted as measures of ethno-religious difference and identity in the Levantine region of the Near East. In covering this ground we will make two detours, one into the range of factors that condition swine husbandry in the Near East, the other into the social and historical concept of cultural identity.

These digressions are called for because the quality and quantity of our archaeological data has, in a sense, run ahead of the theoretical tools we have employed in their interpretation. Much of the current literature about pig use in the lands bordering the eastern Mediterranean either reduces the question of swine husbandry to one of a simple choice motivated by ideology, thus missing the economic, political, and ecological parameters of the problem, or adopts a view of ethnicity/cultural identity long abandoned in the rest of social science. After this restocking of the toolbox for understanding the facts about pig use and avoidance, we will present an historical hypothesis that seeks to account not so much for why the pig was avoided, but, rather, why avoidance came to have such utility in constructing Israelite/Jewish identity.

Conceptualizing Food Production and Culinary Systems

> "And the pig, because it divides the hoof but does not chew the cud, is unclean for you. You shall not eat their meat, and you shall not touch their carcases."
>
> Deuteronomy 14:8

> "The pig, for even though it has divided hoofs and is cleft-footed, it does not chew the cud; it is unclean for you. Of their flesh you shall not eat, and their carcases you shall not touch; they are unclean for you."[2]
>
> Leviticus 11:7–8

A century of zooarchaeological research has taught us that interpretation of the abundances of the various taxa found in ancient sites is a process filled with ambiguity. This ambiguity is driven by the "curse of equifinality," the oft-remarked fact that many archaeological patterns seem to be equally well explained by a variety of causes and that understanding requires the teasing apart of many lines of evidence (Lyman 1994). Despite this situation, the two biblical texts which, for the Israelite followers of Yahweh, prohibit the consumption of pork or even contact with swine are so unambiguous in their terms that a radical simplification in developing archaeological interpretations of pig bone abundances from Iron Age[3] Levantine sites would seem to be in order. In that culture-historical context if no other, pigs, or more properly the absence of pigs, should constitute a mark of cultural identity.

As an example, Israel Finkelstein (1996:206) recently wrote, "food taboos, more precisely pig taboos, are emerging as the main, if not only avenue that can shed light on ethnic boundaries in the Iron I. Specifically, this may be the most valuable tool for the study of ethnicity of a given, single Iron I site." Not to unfairly characterize this position, Finkelstein (ibid.) does note an initial statement of our work (Hesse 1990) which summarized some of the evidence for a number of alternative cultural processes—"pig principles"—that might have encouraged or discouraged the exploitation of swine in the Near East. However, in moving to connect pig avoidance with one form of social identity, Israelite ethnicity, he depreciates and consequently ignores the significance of these alternatives. We suggest that this jump to equate an absence of pigs with the presence of a particular people is unwarranted until the potential of these alternative principles as a system of explanation within which cultural identity is manufactured is exhausted.

Pig use and avoidance in the Near East was (and is) clearly enmeshed in a web of choices and constraints. Placed in a wider theoretical context, the decision to eat or not eat pork is just one example of a knotty problem in historical social science, one that is global in range: how do we account for the pace of change in culinary choices? An often asserted principle is that foodways, compared to most norms, are conservative elements of culture. Since food is a resource that must be ingested almost daily, most often in a social context, and is extraordinarily variable in its preparation and presentation, it is an obvious target for loading with symbolic value (Mintz 1996).[4] The fact that we learn our society's foodways almost from birth makes our culinary experiences intimate reminders of our families and communities, and markers of significant events—powerful elements in our construction of ourselves. Therefore, the sheer weight of meaning that is piled on foods should give them a kind of cultural inertia. Culinary habits should change gradually, incrementally, and at the social margins, where innovation can pass less marked. Often they do. Yet it is obvious that foodways do sometimes change rapidly.

The American explosion in "drive-through fast food" emporia, the appearance of McDonald's in Moscow and Tel Aviv, the replacement of the "family dinner" with small microwaveable meals taken from the refrigerator by each member of the household, and the manias for oats, the oils of certain fish, "blackened" food, and now (and particularly topical here) kosher foods, all defy the notion that foodways should and do evolve slowly. How should we conceptualize the reality? Mintz provides a powerful model. He suggests that foodways evolve within a structure of two kinds of meaning:

> The daily life conditions of consumption have to do with what I called *inside* meaning; the environing economic, social and political (even military) conditions with *outside* meaning. *Inside* meaning arises when the changes connected with *outside* meaning are already under way. These grand changes ultimately set the outer boundaries for determining hours of work, places of work, mealtimes, buying power, child care, spacing of leisure and the arrangement of time in relation to the expenditure of human energy. In spite of their significance for everyday life, they originate outside that sphere and on a wholly different level of social action. In consequence of these changes, however, individuals, families and social groups must busily integrate what are newly acquired behaviors into daily or weekly practice, thereby turning the unfamiliar into the familiar, imparting additional significance at the most humble levels. (Mintz 1996:20; italics in original)

Mintz (1996:28–29) links these two kinds of meaning to the layers of power in human situations as developed by Wolf (1990). At the core of social situations is the individual capacity through personal charisma or power of persuasion to alter or direct events. The scope of these personal possibilities is constrained by the tactical or organizational capacity of various social units to set the terms of action. Finally, tactical decisions are constrained within the structural power of the regional or global system that establishes the economic and political relationships within and among communities. For our purposes, this perspective suggests that understanding the history of swine husbandry and pork consumption in the Levant will be a task not just of reviewing the list of factors that are related to Levantine swine and applying them to specific cases, but of seeing them as combinations of inside and outside meaning as they found expression at each of the transformational moments in the history of pig use. Perhaps it will turn out that the early Israelites, as Finkelstein suggests, constructed their identity partially through the avoidance of pork. To show that, it will be necessary to establish the structural and tactical constraints of the cultural system within which that choice was both possible and salient.

Since much of the literature on pig use and abuse tends to be mono-causal in outlook, we provide below an outline of the wide range of political, economic, social, and religious factors that, acting alone and in concert, have been suggested to affect the abundance of pigs in ancient Near Eastern husbandry systems (see also Hesse 1995; Hesse and Wapnish 1997). We restrict ourselves to those notions that have been evaluated with archaeological data[5] and provide a summary of the various factors, expressing each as a dichotomous variable related to the animal's management or apparent valuation. Our intent is to create a behavioral and ideological matrix for pork production and avoidance within which we can explore the context of the emergence of the pig as a highly charged symbol central to the process of ethnogenesis in Palestine. Our variables describe processes that operate at both of Mintz's levels of meaning, inside and outside, as well as at one or another of all three temporal frames suggested by the *Annales* historians: the long-term geographic or environmental, the medium-term socio-economic, and the short-term sociopolitical. They thus present a structure that can potentially "balance the fleeting event and the persistent process in an unitary socio-historical account" (Knapp 1992:6).

"Pig Principles"

The "Pig Principles" include:

1. *Wet or Dry:* The successful husbandry of pigs demands water in amounts greater than that required by other barnyard stock, a structural meaning. Caroline Grigson (1987, 1995) suggests that annual rainfall levels of approximately 300–350 mm may be required unless substantial infrastructural investments are made (see the discussion by Michael Toplyn [1994:617–663] of Joel Klenck's unpublished studies of archaeofaunal material from the Negev).

2. *Sedentary or Mobile:* Pigs are hard to herd (Harris 1985), though it is not impossible, as is clearly shown by the use of "drover pigs" to support de Soto's sixteenth century expedition through the American southeast. The question is largely: How much labor will the group be willing to devote to this part of the pastoral effort? The answer is usually so little that the adoption of pig husbandry serves to brake any nomadic or migratory aspirations (Ingold 1987) held by a human society. On the other hand, nomadic communities could receive pork in exchange from settled communities. However, no nomad sites in the region have produced significant numbers of domestic pig remains.

3. *Recyclers or Pests:* The enthusiasm pigs have for waste products found at human habitations can be channeled into creating a low-cost urban husbandry, as well as a sanitation service (Grigson 1995). On the other hand, the amount of plant "waste" that different agricultural regimes yield varies greatly, and not all of it is suitable as pig fodder (Preston et al. 1985). Moreover, pigs can be quite destructive in their search for food. So the attractiveness and viability of casual swine management in a city or town is not certain, but rather conditioned by the agropastoral system which engulfs it and the architecture of the settlement in which it is practiced. This is a particularly clear example of the nature of the relationship between Mintz's inside meaning (using pigs as vacuum cleaners) and outside meaning (at the structural level—the agricultural system; and at the tactical level—the residential pattern). The discovery of an ancient pig sty would be as exciting an event as the successful resolution of the search for ancient Levantine stables.

4. *Initial Strategy or Mature System:* Pamela Crabtree (1989) provides another example of the kind of inside-outside meaning structure Mintz advocates for tackling food production issues. She has argued that the rapid reproductive rate of pigs makes swine husbandry attractive to immigrants who are creating new agro-pastoral settlements in their adopted (or conquered) land. This advantage is gradually lost as the economy matures and the stock of animals more capable of producing secondary products rises. At that point cattle, sheep, and goats become more common. This model actually combines elements of two of the following pig principles, nos. 5 and 7.

5. *Low Intensity or High Intensity Agriculture:* In an important article on the historical variations in pork production found in ancient Egypt (particularly the Old Kingdom period), Richard Redding (1991) suggested that the intensification of grain production was accompanied by the shift from a sheep/pig husbandry to one focused on cattle and goats, a transformation required by the changes in land use in the new system. Since pressure for increased production probably originated at the highest levels of the Egyptian state, Redding's proposal can be classified as a structural element in the construction of outside meaning.

6. *High Intensity Agriculture or Carnivorous Pastoralism:* When the preference for meat as a source of food rises to a sufficiently high level, the market reward encourages specialized "carnivorous

pastoralism" (Ingold 1980), even given the inefficiencies of this mode of production compared to those focused on dairy products or fiber. At some price threshold, pigs can re-enter the agro-pastoral system since the profit margin associated with meat then exceeds the value assigned secondary products available from the bovids. It is no accident that chickens and pigs, both species capable of high meat yield, enter (re-enter) the urban market in the southern Levant together during the economic and demographic boom of the Hellenistic and Roman periods.

7. *Independence or Centralization:* Pigs produce much protein rapidly. However, in technologically simple systems, the energy that swine yield is not easily transported to market or captured by tributary systems. These tactical or organizational constraints have led to the idea that pig husbandry is a useful rural subsistence strategy, one that permits satellite communities to emphasize their domestic, non-market-based modes of production and, in so doing, maintain a degree of independence from those centers in the political economy which seek to control them (Diener and Robkin 1978). Evidence of these relationships has been developed for both Early and Middle Bronze Age contexts in the Levant (Hesse 1995).

8. *High Class or Low Class:* Howard Hecker (1982, 1984), in his review of the use of swine in ancient Egypt and, in particular, his study of the Tell el-Amarna faunal remains, observed that both the production and consumption of pork was associated with lower or working class status, both in the textual record and in samples from archaeological remains. This association appears to be duplicated in Mesopotamian texts and archaeological remains as well. McGuire Gibson (pers. comm.) notes that in a Mesopotamian urban site, debris associated with labor gangs is rich with pig bones, while contemporary residential garbage from other sectors of the site is not. He offers in explanation what Mintz would classify as a tactical constraint, that the abundance of pork is a rational response to the needs of administrative officials to feed workers efficiently.

9. *Ritual or Secular:* The use of pigs in religious rituals of the ancient Near East has been remarked by many authorities (de Vaux 1958; Houston 1993; Milgrom 1991). This conclusion, based mostly on textual evidence, is not supported by the archaeological evidence. Particularly striking is the disassociation between pigs and "sacred space" at the tiny Bronze Age site of Tell el-Hayyat in the central Jordan Valley (Mary Metzger, pers. comm.). What we are missing, however, is a sense of the tactical and structural constraints that surrounded Levantine centrally organized sacrificial systems.

The "Meaning" of Pigs

A key point that we are trying to make with this listing is that, taken as a whole, many ancient Middle Eastern peoples associated pigs with the values and experiences of *wet, sedentary, initial, simple, non-market, resistance, meat, rural, urban, low class,* and *secular*—a complex language in swine. When applied to the archaeological record it turns out that rarely is only one of these principles relevant in a particular place and time. For instance, during the Middle Bronze Age in the Levant, principles 2, 7, 8 and 9 would seem to be important to varying degrees. By contrast, in the Philistine cities of the early Iron Age, principles 3, 4, 5, 6 and 8 suggest themselves (Hesse 1995). While no one has yet developed a set of methods which convincingly tease these alternatives apart and assign them specific weight (this largely because of the mono-causal perspective mentioned above), the principles do, however, provide the grammar of inside and outside meaning within which cultural decisions about pig use were and are made.

The History of Pig Use

Despite the ambiguities about particular cases, the available zooarchaeological data are sufficient to look at the historical pattern of pig use from a regional perspective (Hesse 1995; Hesse and Wapnish 1997). Examined at that scale, the outcome of the interaction of the pig principles with myriad local conditions is a sequence of dramatic oscillations. First we see a long-term decrease in the utilization of the animal. Pig production drops from a peak in Neolithic, Chalcolithic, and Early Bronze Age times to a low in the later phases of the Iron Age when precious few communities at all husbanded the animal. This long slide was followed by a sharp rebound in Hellenistic, Roman, and Byzantine times. While pockets of this renewed pig production did persist through the Islamic periods, the subsequent picture over all is one of re-decline. It is against this historical pattern that the emergence of the prohibition on pork should be seen.

Interpretations of Pig Remains

> When I was in Granada, Spain, at a center for research on the history of Spanish food called El Molino, I was told Jewish cooking was one of the major influences there. I asked for an example and was given "the way of roasting pig." In response to my disbelief, the mustachioed gastronome explained that the Marranos cooked suckling pig in

the way they had cooked baby lamb. Converted Jews made a point of cooking pork to protect themselves from charges by the Inquisition of continuing to practice their old religion. Rabbi Abraham Levy, head of the Sephardi congregation in Britain told me that he gets a number of inquiries from people in Spain and Portugal curious about their ancestry who claim that it was the practice for generations in their families to hang hams outside the door. That was one of the ways Jews tried to prove their abandonment of their faith. (Roden 1996:386)

From the perspective of a casual outside observer, the food rules, or the laws of *kashrut*, as represented by the small icons of rabbinic approval printed on food packaging and the Hebrew letters painted on the windows of kosher butcher shops are among the most visible symbols of a Jewish community. Of the complex set of rules about slaughter, food preparation, the permissible combinations of dishes in a meal, and the religious specialists who supervise the processes that actually make up the system (Cohn 1981), no element is more widely known than the prohibition of pork. The equation of Jewish identity with the avoidance of swine as prescribed in Deuteronomy 14 and Leviticus 11 would appear to be straightforward and unambiguous. Since the prohibition of pork is also specified in the Koran, this relationship applies to Muslims as well. On the other hand, the traditions of Christians, following the innovations of Paul, largely relieve them of the need to express religious conviction through mandated behavior related to foods. Thus, deposits created by the activity of Christians are expected to contain the remains of pigs.

There are numerous examples of the use of these simple cultural identity rules in interpreting archaeological materials. For instance, Horwitz, Tchernov, and Dar (1990) ascribe the increase in pig remains in Medieval levels at the small community of Khirbet Sumaqa on Mount Carmel to the establishment of Christian communities in the region. The fact that earlier layers contain the remains of a synagogue and associated structures is offered to explain the absence of the animal's bones in those deposits. Similarly, Clason (1995:98) interprets a small number of pig bones at Ta'as, a Byzantine and Islamic period site in Syria, to the "users of the Christian chapel in Pit A." One of us, as well, has engaged in this kind of speculative identification. At the site of Caesarea, a large port city on the Mediterranean coast south of Haifa, Hesse and Claudia Christian found large numbers of pig bones in a sample of Fatimid date, a finding we explained by arguing that in the cities they controlled, Muslim rulers allowed Christians (as well as Jews) considerable autonomy of action (Christian and Hesse 1995).

While these particular interpretations may appear reasonable given the textual evidence that supports the three pro-pig people and anti-pig people equations in Roman, Byzantine, and Islamic settings, as a way of proceeding to examine earlier archaeological evidence they veer sharply in the direction of assuming the very relationship that one would like to demonstrate. This problem has two aspects, which Eric Wolf characterizes well in his comment, "By endowing nations, societies, or cultures with the qualities of internally homogeneous and externally distinctive and bounded objects, we create a model of the world as a global pool hall in which the entities spin off each other like so many hard and round billiard balls" (1982:6). First, the impenetrable surface of the billiard ball implies that cultural identity may be a process of group self-definition largely unaffected by external conditions. The symbols which come to express who a people are grow from relationships found solely within their inner cultural world. They are deployed in, not shaped by, interaction. In fact, in the literature which has grown up around the pig prohibition, this is a common stance, one adopted by those who examine only the biblical texts in their search for understanding. For these scholars, the rationale behind the passages laying out the pig prohibition is a matter of religious principle transmogrified by the needs and objectives of the priestly editors who selected, shaped, and eventually canonized the texts. While this approach can produce valuable results such as E.B. Firmage's (1990) elegant deconstruction of the Levitical and Deuteronomic lists of prescribed and proscribed animals as a key for identification intended to help the faithful avoid error, rather than a zoological classification, it does not account for why specific aspects of the religious code come to have such profound behavioral salience.

Second, the permanence of the billiard ball suggests that, once established, the relationship of material culture signal and people may be both retrojected into the past and projected into the future. The salience of a symbol now is taken as a guarantee of its significance then. To translate to the case at hand, this principle can be stated as: Jews, Muslims, and Christians each seek to continuously reproduce their corporate cultural and social identities through the expression of their individually reasoned attitudes toward pork and pigs. A primordial approach to ethnicity/cultural identity such as this provides little scope for testing with archaeological materials.

One profitable way to restructure understanding of the nature of cultural identity is to focus on differences rather than similarities.[6] A key contribution of Fredrik Barth's (1969) work is his emphasis on boundaries. Not only did a group have to recognize its own set of symbols, neighbors had to read and respond to them as well.

Identity arises from interaction. Such an approach has been explicitly applied in identifying an emergent Israelite ethno-religious identity from the distribution of pig remains we were able to report for the early Iron Age in Canaan (Hesse 1986). Lawrence Stager (1995:344) writes of the contrast between the use of pigs by Philistines on the southern coastal plain of Palestine and the absence of the animal in contemporaneous sites in the Hill Country just to the east:

> There can be little doubt that these differences in pig production and consumption have more to do with culture than ecology. The Mycenaeans and later Greeks had a positive attitude towards swine and a preference for pork in the diet. The Philistines brought that preference with them to Canaan in the 12th century. Probably at that time during the biblical 'Period of Judges,' the pork taboo developed among the Israelites as they forged their identity partly in contrast to their Philistine neighbors. Thus during Iron Age I the pig becomes a distinctive cultural marker, just as circumcision was, between Philistine and Israelite.

The difficulty with this straightforward approach has to do with the perspectives of the cultural participants. Malcolm Chapman (1993:23) cogently observes, "most ethnic terminologies...turn out to consist of a complex intermixture of naming by self and naming by others... Surprisingly large numbers of people live in categories which are of someone else's devising." He provides the sobering example of how the interaction of three self-aware social units can produce six 'ethnic groups' if each sees the other two as a single 'other'. He further observes, "Long and careful study is not...something that different cultures are commonly disposed to accord to one another. A degree of self-congratulatory ignorance about other cultures seems to be the *normal* state of humankind" (ibid.:40, emphasis in original).

The problem in the case of the proposed Philistine-Israelite swine-based boundary as developed by Stager is that the perspective of the pig users as they look out at their social surroundings is not a simple flipped reflection of that seen by the pig avoiders as they look back. While the Israelites may have lumped all the Philistines into one undifferentiated group of (uncircumcised) swine lovers, in fact, pig use in the early Iron Age in the southern Levant is found in some Philistine communities—notably the port city of Ashkelon, and Ekron and Timna, two sites situated in the rolling hills of the Shephelah (the territory between the coastal plain and the Hill Country)—but not all of them—notably Qasile, a site located near the coast in modern Tel Aviv (Hesse 1990, 1995; Hesse and Wapnish 1997). Clearly there were some Philistine "apostates" failing to consume their allotments of pork.

Further, there is a decrease in the remains of swine as one moves to the margins of Philistine territory. By the time one reaches Beth-Shemesh at the edge of the central Hill Country southwest of Jerusalem, pig remains are essentially absent despite the presence of some distinctive Philistine pottery. This fact has been used to support the identification of the site as non-Philistine and to treat the pottery as a result of trade (Zvi Lederman, pers. comm.). At Tell Jemmeh, just inland from Gaza to the south, pigs are also nearly absent in levels where a large Philistine kiln was found (Van Beek 1993). The anomaly of this non-association of Philistine and pig is underlined by the abundance of the animal in Middle Bronze Age deposits at the same site (Wapnish and Hesse 1988), suggesting that we cannot explain away this particular absence simply on environmental grounds. Variations in pig abundance within single Philistine sites suggest that internal social distinctions may have shaped access to pork. Thus there is clearly a spatial texture to the exploitation of the animal by these Sea People importers of Mycenean/Greek culture to their new Levantine homeland.

Moreover, pig use by the Philistines is restricted to only the initial phases of their occupation, perhaps the first century or two after their arrival ca. 1175 B.C.E. This observation, initially made based on the first year's work at the coastal Philistine center of Ashkelon (Hesse n.d.b), has been reinforced by the evidence accumulated during subsequent years. The Philistine community smashed by Nebuchadnezzar in 604 B.C.E. consumed hardly any pork. If the sample from the huge fill underlying part of the area destroyed by the Babylonian king is any guide, this was a culinary tradition they had been following for several hundred years while interacting and maintaining a cultural boundary with the presumably pig-hating Judaeans to their north and east who, ironically enough, despised their rivals for their supposed love of pig.

Turning ourselves around and looking out towards the Israelites from the perspective of the Philistines (and assuming that a unitary view is appropriate!), pig avoidance seems to have been ubiquitous. Not only were the "Jews" through their long history both as a United Monarchy and as the divided kingdoms of Israel (the northern polity which was finally conquered by the Assyrians at the end of the eighth century B.C.E.) and Judah (the southern region which survived until the Babylonian victory) avoiding the animal, so too were almost all other societies in southern/southwest Asia during the Iron Age.

In the regions close to Philistia, pig avoidance is universal. While the lack of pigs is noticeable at sites located in the traditional early Israelite sphere—such as Izbet Sartah, situated on the fringe of the coastal plain to the east of Philistia; Ai, Raddana and Shiloh, further removed in the central Hill Country; Lachish, located in the southern reaches of the Hill Country; and Dan, situated in the Upper Galilee—it extends into adjacent territory as well. In fact, the pattern reported above for Tell Jemmeh is duplicated at Ein Hagit, a small site of uncertain cultural affiliation located in the Carmel range north of the border of Philistia. There we have an earlier Middle Bronze Age dependence on pigs replaced with a husbandry excluding swine in the first phases of the Iron Age after a Late Bronze Age hiatus of occupation (Hesse n.d.a). Is the absence of pigs enough to justify including this small settlement in the Israelite cultural sphere despite its location and nondiagnostic artifactual/architectural assemblage? Rigorously applied, such a procedure of equating the absence of pigs with cultural identity would lead to a remarkable (and preposterous) expansion of early Israelite hegemony, since in the Greater Levant, including what we know of Syria, Lebanon, and Jordan, only the site of Hesban in Jordan has produced more than a minuscule amount of pig, and here it is only ~ 5% (LaBianca and von den Driesch 1995:72).[7] As mentioned above, it is not until the Hellenistic period that pigs are widely found in Levantine sites and any remnant Philistines would have seen communities embracing their ancestral (and long discarded) culinary tradition.

It is necessary to move much farther afield, to Anatolia, to find significant examples of Iron Age swine management. Here it is important to note that even in cases where pig is present, bias against the animal is reported in the same breath, usually understood in terms of Pig Principle 9: Ritual or Secular. In Marian Fabiš's (1995) report on a sample from Troy, for instance, the absence of pigs from one area of the site is explained by the area's sacred character. Similarly, Joris Peters (1993) explains the absence of pig remains in a seventh–fifth century B.C.E. sacred area at another site (Zeytin Tepe) in western Anatolia with a quotation from Pausanias in which it is asserted that pigs are not a suitable sacrifice for Aphrodite.

What is evident is that the distribution of pig avoidance/non-use far exceeds even the most generous limits for the Israelite polity. As the Philistines would have seen it if they were in fact pig lovers, the entirety of their surrounding world of cities, states, and empires—some allies, some enemies—was undifferentiated, at least in terms of cuisine, and united in the avoidance of swine.[8] Clearly a wider range of causal factors than ethnic identity has to be invoked to explain all these pig absences.

An additional way to approach the plastic and multifaceted nature of cultural identity is with the concepts of assimilation and acculturation. One example is provided by Jako Weinstock, based on his work at Carthage. He writes (1995:115):

> How can this gradual increase in pork consumption be explained? From historical sources we know that people belonging to a number of non-Phoenician/Punic ethnic groups were living in the city, Greeks, Etruscans, Sicilians, and Anatolians among others. Assuming culture-caused differences in the degree of pork consumption, one could postulate that the number of foreigners in the city increased as the city developed and grew. An alternative, but related explanation would be the following: pork-loving people from non-Punic origin lived spatially segregated from the Punic inhabitants. The former ate larger amounts of pork than the latter, resulting in differences in the spatial distribution of pig remains. The gradual increase in pork consumption could then be interpreted as resulting from "acculturation" of the Semitic (Phoenician) people.

Weinstock's approach is valuable since it (1) emphasizes the negotiated nature of culinary choices in a multicultural setting, (2) predicts that cultural identity may have spatial expression in a complex urban environment, and (3) suggests that the shifting patterns may have archaeological visibility. It serves as a useful empirical expression of the way social science has come to understand the notion of cultural identity. However, it does not develop the wider eco-political framework in which acculturation was possible.

Recent treatments of cultural identity (e.g., Eriksen 1993; Graves-Brown, Jones, and Gamble 1996; Shennan 1989) reject the notion of rigid symbolic content in favor of a more historically contingent approach. The quotation which heads this section reflects the shifting significance of dietary observance both in the minds of Jews and those among whom they lived. In the case of *kashrut* today, both in Israel and in the diaspora, what is not so obvious to the outsider is that, as assimilationist pressures and culinary innovations have their impact, debate rages over how the laws are to be observed or even whether it is essential to follow them. In fact, since the advent of modernity in the eighteenth and nineteenth centuries, the basis of Jewish identity has undergone renegotiation to a degree not seen since the first centuries of the common era. Not only is the form of practices like the food laws, Sabbath observance, and circumcision contested, but even the tradition of tracing descent solely through the mother. The lesson to be drawn from the modern Jewish experience is clear. The construction of identity is, as the Comaroffs (1992) assert, both strategic and contextual. Components of the tradition are selected for the imposition of cultural meaning in response to surrounding historical conditions based on the lessons of experience, perceptions of advantage, and social goals. The formation of ethnic identity, in particular, is fostered by social stratification as a mechanism for protecting and enhancing group status. In this reformulation of the concept, the question of the association of pig avoidance and cultural identity is transformed from an equation to one of a context. No one can doubt that the Israelites were not using pork in their cuisine for centuries—the empirical evidence is simply overwhelming—the problem is to determine when that behavior was mobilized to socio-religious and political effect.

The Return from Exile and the Food Laws

We are reminded by Douglas (1966) that the textual evidence for Israelite pig disdain is sparse. Nevertheless, a considerable contrast can be seen when passages in the Hebrew Bible reflective of the period prior to the exile to Babylon in the sixth century B.C.E. are contrasted with those passages which describe Jewish life in the centuries immediately after the exiles' return to their homeland.

The exile was a foundational event in Jewish history. In his conquest of Syria-Palestine in 586 B.C.E., the Babylonian king Nebuchadnezzar forced many Jews to leave their homeland and resettle in Babylonia. These forced migrants, having experienced a devastating defeat that included the destruction of the central edifice in their religious topography, the temple in Jerusalem, and having been ripped from their homeland, had to confront substantial issues of religious and cultural identity in order to preserve any sense of cohesion in their new communities. Clearly the traditions and beliefs that had sustained them for centuries had to be questioned by those in exile, and new ways of understanding their situation and maintaining their identity proposed. Then, a half century later (ca. 538 B.C.E.), after Achaemenid troops had overthrown Babylon and taken over its dominions, the Persian king Cyrus permitted the Jews to return to their homeland. Despite the welcome opportunity to reclaim the land of Israel, the process through which the Jewish community was re-established was neither rapid nor uncontested. A key figure, at least as expressed in the Hebrew Bible, was Ezra, a priestly leader whose policies instituted sometime in the fifth century B.C.E. profoundly altered the basis of official Jewish life. What seems clear is that the reconstituted Jewish community, an amalgam of returnees and those who had not been forced to leave, was a much changed national, religious, and cultural entity.

It is our contention that the differences between the biblical passages reflective of pre-exilic and post-exilic date signal that the fundamental basis of identity for the Jews was changed by the diaspora experience. Under the new cultural and religious rules the salience of the pig as a common barnyard animal and pork as a component of the cuisine was reinforced and prepared for deployment as a marker of cultural identity in challenges yet to come.

To the legislation in Leviticus 11 and Deuteronomy 14, only two passages in the Hebrew Bible decry the consumption of pork:

> a people who provoke me to my face continually,
> sacrificing in gardens and offering incense on bricks;
> who sit inside tombs, and spend the night in secret places;
> who eat swine's flesh, with broth of abominable things in their vessels
>
> Isaiah 65:3–4

> Those who sanctify and purify themselves to go into the gardens,
> following the one in the center, eating the flesh of pigs, vermin and
> rodents, shall come to an end together, says the Lord.
>
> Isaiah 66:17

If citation rate is a measure of how common or serious a violation was considered, not many people were eating pork and there was not much concern about those who did.

An additional problem is the date of these texts. Some critics treat the anti-pig bias as a part of the earliest traditions of the Israelites, even if the date of the actual legislation can only be placed in the seventh or eighth century B.C.E. on linguistic grounds (Milgrom 1991).[9] They argue that the injunctions were among the foundational religious rules of the Israelites, ones that were in place by the early centuries of the Iron Age, a position essentially adopted by Finkelstein in the article mentioned above. Others suggest that the biblical texts we have today were assembled only in the last centuries B.C.E. and only reflect conditions at that time (e.g., Davies 1992; Thompson 1992).

Whatever the outcome of the philological debate over the redaction of the Taanakh, these passages from Isaiah give some insight into the organization and diversity of Israelite religious practice and the social structure within which it was expressed (at least as it was remembered) in the century before the exile. In both cases, the violators of the law are seen to creep off into a private place to conduct their abominations, but, though condemned, are still in some sense part of the community.[10] These incidents correspond to the portrait of Iron Age Israelite religion constructed from archaeological sources by Holladay (1987:281). He sees a contrast and tension between "an officially established, hierarchically organized state religion...operative in close coordination with the state's political apparatus" and "cultic remains...totally isolated from the life of the official shrines and sanctuaries...best explained as popular phenomena, probably dependent upon traditions of folk religion stretching back into the Bronze ages..." The picture is one of a people with a diversity of beliefs who find their shared identity in values that incorporate but transcend the practices of official religion. A pivotal aspect of the root of that broader identity was shared possession of the Land.[11] From their perspective, the Israelites were living in the land that Yahweh had promised them, and they could expect to transmit that physical-religious heritage to future generations through descent and inheritance. Cultural identity was at least as much, and probably more, a matter of residence than one of practice.

The citations from the Hebrew Bible contrast in both form and substance with the sacred demand for pig avoidance found in the apocryphal/deuterocanonical books of Judith and particularly Maccabees, texts presumptively later than the biblical passages.[12] These stories record traditions set in periods from the Neo-Babylonian to the Hellenistic, though in the form we have them the texts are probably Hellenistic. In the book of Judith, the heroine is entertained by the Assyrian general Holofernes. In her conversation with him she reports that her nation has drawn God's anger and withdrawal of protection because "they have planned to kill their livestock and have determined to use all that God by his laws has forbidden them to eat" (Judith 11:12). Later Holofernes invites Judith to join him in eating a meal prepared by his servants, an offer she refuses, saying, "I cannot partake of them, or it will be an offense; but I will have enough with the things I brought with me" (Judith 12:2). More dramatic and specifically related to the consumption of pork are the passages in 2 Maccabees, here reflecting conditions in the last centuries before the Common Era. For example,

> Eleazar, one of the scribes in high position, a man now advanced in age and of noble presence, was being forced to open his mouth to eat swine's flesh. But he welcoming death with honor rather than life with pollution, went up to the rack of his own accord, spitting out the flesh, as all who ought to go who have the courage to refuse things that it is not right to taste, even for the natural love of life.
> 2 Maccabees 5:18–20

> It happened also that seven brothers and their mother were arrested and were being compelled by the king, under torture with whips and thongs, to partake of unlawful swine's flesh.
> 2 Maccabees 7:1

We can identify three points of comparison between these two sets of texts that are especially significant. First, in the verses in Isaiah the focus is on the consumption of pork, not its avoidance as we see in the later texts. Second, in contrast to the passages in Isaiah where the setting of purported pig use is private and general, each episode of obeying the food laws in Judith and Maccabees is placed in a concrete setting of apparent historical character. Third, the act of consuming pork described in Isaiah is only observed by God, whose fury is provoked. In Judith and Maccabees the refusal to violate the rules is observed and reacted to by agents of other societies. The avoidance of pork has become a central act of identity in the secular as well as the religious dimension, one to be resisted even if it means a horrible death.[13]

What has happened in the intervening years between the legislation of Deuteronomy and Leviticus—texts perhaps as old as the seventh or eighth century B.C.E. in which pigs receive hardly more attention than the hyrax—the two mentions in Isaiah, and these ghastly records of suffering in 2 Maccabees? Two events are pivotal. First, as mentioned above, was the traumatic experience of defeat at the hands of the Babylonians and removal to Mesopotamia and the Nile Valley. The second was the challenge of Hellenism.

Exile for defeated communities was not an innovation under the Babylonians. Assyrian conquerors had been using the policy for centuries. What was different was the scale of the process after the conquest of Jerusalem. While many Judaeans did remain in Judah, a large number were detached from the primary source of identity, the Land, and forced to find a new basis on which to construct their identity. Clearly, many assimilated. However, some sought to maintain a community in exile. Ahlström (1986:109–110) characterizes the response of the latter group this way:

> Faced with the possible loss of identity, the religious leaders of the exiles make the bold and rad-

ical assertion that Yahweh was not bound to a state or a particular territory like other deities. Ezekiel, for instance, asserted that Yahweh had followed his people into exile and thus, could still be worshiped, even if the nation and its temple no longer existed... One might say that the "congregational" idea of religion was born in the Exile to help the deported Judahites rationalize their circumstances.

Proper observance of religious law came to replace the now broken ties to the land that had previously held the nation together. Ezekiel was particularly forceful about this. As phrased by Miller and Hayes (1986:434):

> Undoubtedly many Judeans, like most of the Israelites exiled by the Assyrians, assimilated the culture so thoroughly as to lose their Jewish identity. Prophets as radically different as Ezekiel and Second Isaiah (the author of Isa. 40–55) wrote and preached during the exile. Ezekiel, with his priestly orientation, especially condemned his contemporaries for their abominations and impurities, their adherence to idolatry, and their syncrestic worship of other gods in addition to Yahweh.

It was a group whose identity was founded on this basis who returned from exile to establish a Jewish entity centered in Jerusalem and rebuild their temple under Persian auspices. There they found themselves surrounded by descendants of the Judaeans who had not left. These individuals had ties to the community that were still based on land, though their religious practices may have relaxed or diverged widely from one another, as well as from those who had been taken into exile, a process made easier by the absence of central religious authority in Jerusalem.

The laws instituted by two early leaders, Ezra and Nehemiah, reflect the seriousness of the conflict between the natives and the returnees. Detached from the Land and Jerusalem, a profound part of the exilic experience for the Jews was their inability to have the priestly class perform the required sacrifices at the temple. A component of their solution to this problem was to emphasize the significance of proper behavior as established by biblical texts. Thus, though they were busy rebuilding the temple in Jerusalem, the returning community advocated a strict adherence to deuteronomic law, even insisting that marriage to non-Jews was an abomination, Ezra going so far as to demand that existing marriages to non-Jews be dissolved. An additional and crucial aspect of Ezra's innovations was the practice of public Torah reading, an institution that had the effect of democratizing the religion and establishing the obligations of lay adherents to the religion in the same frame as the priestly class. These changes created ideological fissures between the returnees, whose activities were supported by the Persian authorities, and those communities who had not been forced into exile. Ahlström (1993:886) summarizes the process this way:

> With Ezra's reform the term 'Israel' took on a new and narrower meaning. It became purely a religious idea. The geographical and political concept of Israel had been replaced by a concept that referred to those of the *gôlâ* party who embraced the law of Ezra and who could establish their descent. Membership in this society was no longer politically or territorially dependent. It was based on a religious ideology that excluded other worshipers of Yahweh.

The nature of cultural identity for Jews had been radically transformed. A strategy for surviving exile had been transplanted to the homeland, a process perhaps aided by the persistent worry that, since the land had been lost once, it could be lost again. Community was now based on strict adherence to a code of behavior. Plurality in religious belief could no longer be sheltered under an umbrella of identity based on residence and inheritance.

It was with this divisive calculus of ethno-religious identity that Jews faced the second crucial event, the arrival of Alexander the Great and the challenge of Hellenistic culture, one component of which was the love of pork.

Conclusion

In previous work we have argued that beginning at least as early as the Bronze Age, the Near Eastern pig was burdened with potential and contradictory meanings. Depending on when and where one looks, dichotomies in meaning constructed on ecological, social, political, economic, and religious grounds structured the use of the animal and produced a matrix of symbolic possibilities available for mobilization. We suspect that the Israelites absorbed this complex of attitudes in a way not dissimilar from their numerous neighbors. There was nothing particularly unusual about the avoidance of pigs in certain sectors of most Near Eastern societies; it doubtless was a strategy reinforced by economic realities, and so Israelites too followed the standard ways of the region, at least as best we can judge from the archaeological record. What was needed to transform pig avoidance into pig hate and a defining element of cultural identity for the Israelites was first, the shift in the basis of community from possession of land to religious practice, and second, the threat of assimilation into a politically dominant society with an economy which rewarded the production of meat at a sufficient level to encourage swine husbandry for the urban market.

Acknowledgments

We would like to express our appreciation to the many archaeologists who have entrusted their collections of faunal remains to us: Avram Biran, Amihai Mazar, Lawrence Stager, Israel Finkelstein and David Ussishkin, Zvi Lederman and Shlomo Bunimovitz, Sam Wolff, Trude Dothan and Sy Gitin, Gus Van Beek, and J.P. Dessel and Bonnie Witthoff. These samples provided the empirical basis on which we began to examine the issue of the pig as a marker of identity. Baruch Halpern provided much useful advice and direction on textual matters. We remain responsible for all errors of fact and interpretation that remain.

Notes

1. Mary Douglas presented the major statement of her idealist view of the problem of the food laws in Deuteronomy 14 and Leviticus 11 in her book *Purity and Danger* (1966). A recent presentation of Marvin Harris's cultural materialist view of the prohibition of the pig can be found in his book *Good to Eat* (1985). In his 1991 edition of *Leviticus 1–16,* Jacob Milgrom tackles both the question of the date of the biblical legislation and the issue of the religious context of the rules. The recent book-length treatment by Walter Houston (1993) places the particular case of the pig in the broader context of the evolution of Israelite religious ideas.

2. These and subsequent quotations from biblical and apocryphal texts are taken from *The New Oxford Annotated Bible* (New Revised Standard Version), Oxford University Press, 1991.

3. The Iron Age in the eastern Mediterranean is conventionally dated from ca. 1200 to 586 B.C.E.. In traditional Israelite archaeological history, Iron Age I culminates with the emergence of the United Monarchy of Israel and Judah about 1020 B.C.E. During Iron Age II this unity proves brief; the two polities divide after about a century to follow separate historical trajectories. In 722 B.C.E. the Assyrians complete the conquest of the northern kingdom of Israel. Judah, the southern kingdom, survives until 586 B.C.E. when it falls to the Babylonians and the period of exile begins. In 538 B.C.E. a Persian edict permits Israelites to return, but the pace and scale of this reverse migration is not well understood. In 332 B.C.E. the Hellenistic period begins.

4. Sidney W. Mintz's recent book, *Tasting Food, Tasting Freedom* (1996), is an enormously insightful approach to the political and social structures within which culinary choices are made. It has reoriented our appreciation of the problems we face as zooarchaeologists in tracing the evolution of foodways.

5. Much of the literature on the pig in the ancient Levant treats the problem as one in logic rather than evidence. For instance, it has been suggested that the decline in pig use is linked to changes in forest cover and a restriction in pig habitat. However, this important suggestion has not been followed up, to our knowledge, with a detailed study of the correlation between vegetational patterns and pig bone abundances.

6. See the extensive discussion of ethnicity in archaeology in Emberling 1997.

7. A central question in evaluating the significance of pig bone statistics is How much is enough? We suggest that, given the rapid reproductive rate of pigs compared to other barnyard mammals, any estimate of their presence in an attritional archaeological deposit (Hesse and Wapnish 1985) will overstate their proportional number in the living herd of barnyard animals that supported the site at the time in question. Thus a value of 5% pig in the *thanatocoenosis* probably signals a less than ubiquitous animal in the *biocoenosis* of ancient Hesban.

8. Shlomo Bunimovitz (1990:210–222) has made a useful case for the notion that Philistine identity is in fact more a product of modern analysis than ancient practice.

9. Excellent introductions to the complexities surrounding the emergence of the Hebrew Bible can be found in Friedman 1987 and Gottwald 1985.

10. De Vaux (1958) treats these passages as evidence of a mystery cult in ancient Israel. Ackerman (1992), by contrast, suggests that these practices were part of the popular religion that accompanied the official cult during the Iron Age. In either case, the texts suggest the existence of considerable diversity of practice within the Israelite community.

11. We are indebted to CLAL Scholar Rabbi Brad Hirschfeld who introduced us to the importance of the shifting basis of Israelite identity, the significance of the first diaspora, and the impact of Ezra and Nehemiah on the evolution of Judaism.

12. Fabre-Vassas (1997:6) points out that Mary Douglas early on noted the significance of the text from Maccabees: "As Mary Douglas emphasizes...refusing to eat pork...is...to perform an act of allegiance and fidelity to the ancestral laws, that, along with this taboo have chosen circumcision and respect for the Sabbath as signs of belonging."

13. It is worth observing that this text of Maccabees, in that it concerns mastery of one's self, in this case by refusing to eat pork, is an expression of the "definitive masculine trait in most of the Greek and Latin literary and philosophical texts that survive from antiquity" (Moore and Anderson 1998:250). This thematic tie to the Classical world further emphasizes the distinctiveness of avoidance as a mechanism of cultural identity in the Hellenistic period for Jews as compared to earlier times.

References

Ackerman, S. 1992. *Under Every Green Tree. Popular Religion in Sixth-Century Judah.* Harvard Semitic Monographs No. 46. Harvard Semitic Museum, Cambridge, MA.

Ahlström, G. 1986. *Who Were the Israelites?* Eisenbrauns, Winona Lake, IN.

——— 1993. *The History of Ancient Palestine.* Fortress Press, Philadelphia.

Barth, F. 1969. *Ethnic Groups and Boundaries: The Social Organization of Cultural Difference.* Scandinavian University Press, Oslo.

Bunimovitz, S. 1990. Problems in the "Ethnic" Identification of the Philistine Material Culture. *Tel Aviv* 17:210–222.

Chapman, M. 1993. Social and Behavioral Aspects of Ethnicity. In *Social and Biological Aspects of Ethnicity*, ed. M. Chapman, pp. 1–46. Oxford University Press, Oxford.

Christian, C., and B. Hesse. 1995. A First Look at Fatamid Animal Exploitation: The Sample from Caesarea Vault II. Paper presented at the annual meeting of the American Schools of Oriental Research, Philadelphia, Nov. 19, 1995.

Clason, A.T. 1995. Ta'as, a Late Byzantine, Early Islamic and Ayyubid Site in Northwest Syria. In *Archaeozoology of the Near East II*, ed. H. Buitenhuis and H.-P. Uerpmann, pp. 97–104. Backhuys Publishers, Leiden.

Cohn, Rabbi J. 1981. *The Royal Table. An Outline of the Jewish Dietary Laws.* Feldheim Publishers, Jerusalem.

Comaroff, J., and J. Comaroff. 1992. Of Totemism and Ethnicity. In *Ethnography and the Historical Imagination*, ed. J. Comaroff and J. Comaroff, pp. 49–67. Westview Press, Boulder.

Crabtree, P.J. 1989. Sheep, Horses, Swine and Kine: A Zooarchaeological Perspective on the Anglo-Saxon Settlement of England. *Journal of Field Archaeology* 16(2): 205–213.

Davies, P.R. 1992. *In Search of Ancient Israel.* Journal for the Study of the Old Testament, Supplement Series 148. Sheffield Academic Press, Sheffield, England.

de Vaux, R. 1958. Les sacrifices de porcs en Palestine et dans l'Ancien Orient. *Beihefte zur Zeitschrift für die Alttestamentliche Wissenschaft* 77:250–265.

Diener, P., and E.E. Robkin. 1978. Ecology, Evolution, and the Search for Cultural Origins: The Question of Islamic Pig Prohibition. *Current Anthropology* 19:493–540.

Douglas, Mary. 1966. *Purity and Danger.* Praeger, New York.

Emberling, G. 1997. Ethnicity in Complex Societies: Archaeological Perspectives. *Journal of Archaeological Research* 5(4): 295–344.

Eriksen, T.H. 1993. *Ethnicity and Nationalism: Anthropological Perspectives.* Pluto Press, London.

Fabiš, M. 1995. Animal Bones from the Classical City of Ilion (Troy), Turkey. In *Archaeozoology of the Near East II*, ed. H. Buitenhuis and H.-P. Uerpmann, pp. 105–108. Backhuys Publishers, Leiden.

Fabre-Vassas, C. 1997. *The Singular Beast; Jews, Christians and the Pig.* Columbia University Press, New York.

Finkelstein, I. 1996. Ethnicity and Origin of the Iron I Settlers in the Highlands of Canaan: Can the Real Israel Stand Up? *Biblical Archaeologist* 59(4): 198–212.

Firmage, E.B. 1990. "The Biblical Dietary Laws and the Concept of Holiness. In *Studies in the Pentateuch*, ed. J.A. Emerton, pp. 177–208. Vetus Testamentum Supplement 41. Brill, Leiden.

Friedman, R.E. 1987. *Who Wrote the Bible?* Harper and Row, New York.

Gottwald, N.K. 1985. *The Hebrew Bible. A Socio-Literary Introduction.* Fortress Press, Philadelphia.

Graves-Brown, P., S. Jones, and C. Gamble (eds.). 1996. *Cultural Identity and Archaeology.* Routledge, London.

Grigson, C. 1987. Shiqmim: Pastoralism and Other Aspects of Animal Management in the Chalcolithic of the Northern Negev. In *Shiqmim I*, ed. T.E. Levy, pp. 219–241, 535–546. BAR International Series 356. Oxford.

——— 1995. Plough and Pasture in the Early Economy of the Southern Levant. In *The Archaeology of Society in the Holy Land*, ed. Thomas E. Levy, pp. 246–268. Facts on File, New York.

Harris, M. 1985. *Good to Eat.* Simon and Schuster, New York.

Hecker, H. 1982. A Zooarchaeological Inquiry into Pork Consumption in Egypt from Predynastic to New Kingdom Times. *Journal of the American Research Center in Egypt* 19:59–69.

——— 1984. Preliminary Report on the Faunal Remains from the Workmen's Village. In *Amarna Reports I*, ed. B.J. Kemp, pp. 154–164. The Egypt Exploration Society, London.

Hesse, B. 1986. Animal Use at Tel Miqne-Ekron in the Bronze Age and Iron Age. *Bulletin of the American Schools of Oriental Research* 264:17–27.

——— 1990. Pig Lovers and Pig Haters: Patterns of Palestinian Pork Production. *Journal of Ethnobiology* 10:195–225.

——— 1995. Husbandry, Dietary Taboos and the Bones of the Ancient Near East: Zooarchaeology in the Post-Processual World. In *Methods in the Mediterranean: Historical and Archaeological View on Texts and Archaeology*, ed. D.B. Small, pp. 195–232. Brill, Leiden.

——— n.d.a [1997]. The Mammalian Fauna of Ein Hagit. In *'Atiqot*, ed. S. Wolff. Forthcoming.

——— n.d.b [1999]. The Animal Economy of Iron Age Ashkelon. In *Ashkelon I*, ed. L.E. Stager. Harvard Semitic Museum, Cambridge.

Hesse, B., and P. Wapnish. 1985. *Animal Bone Archeology: From Objectives to Analysis*. Taraxacum, Washington, DC.

——— 1997. Can Pigs Be Used for Ethnic Diagnosis in the Ancient Near East? In *The Archaeology of Israel: Constructing the Past, Interpreting the Present*, ed. N.A. Silberman and D.B. Small, pp. 238–270. Sheffield, Sheffield University Press. Forthcoming.

Holladay, J.S., Jr. 1987. Religion in Israel and Judah Under the Monarchy: An Explicitly Archaeological Approach. In *Ancient Israelite Religion*, ed. P.D. Miller, Jr., P.D. Hanson, and S. D. McBride, pp. 249–299. Fortress Press, Philadelphia.

Horwitz, L.K., E. Tchernov, and S. Dar. 1990. Subsistence and Environment on Mount Carmel in the Roman-Byzantine and Mediaeval Periods: The Evidence from Kh. Sumaqa. *Israel Exploration Journal* 40:287–304.

Houston, W. 1993. *Purity and Monotheism: Clean and Unclean Animals in Biblical Law*. Journal for the Study of the Old Testament, Supplemental Series 140. Sheffield Academic Press, Sheffield, England.

Ingold, T. 1980. *Hunters, Pastoralists and Ranchers*. Cambridge University Press, Cambridge.

——— 1987. Changing Places: Movement and Locality in Hunter-Gatherer and Pastoral Societies. In *The Appropriation of Nature, Essays on Human Ecology and Social Relations* by T. Ingold, pp. 165–197. University of Iowa Press, Iowa City.

Knapp, A.B. 1992. Archaeology and *Annales*: Time, Space and Change. In *Archaeology, Annales, and Ethnohistory*, ed. A.B. Knapp, pp. 1–21. Cambridge University Press, Cambridge.

LaBianca, Ø., and A. von den Driesch (eds.). 1995. *Faunal Remains: Taphonomical and Zooarchaeological Studies of the Animal Remains from Tell Hesban and Vicinity*. Hesban 13. Andrews University Press, Berrien Springs, MI.

Lyman, R.L. 1994. *Vertebrate Taphonomy*. Cambridge University Press, Cambridge.

Milgrom, J. 1991. *Leviticus 1–16*. The Anchor Bible, Vol. 3. Doubleday, New York.

Miller, J.M., and J.H. Hayes. 1986. *A History of Ancient Israel and Judah*. The Westminster Press, Philadelphia.

Mintz, S.W. 1996. *Tasting Food, Tasting Freedom: Excursions into Eating, Culture and the Past*. Beacon Press, Boston.

Moore, S.D., and J.C. Anderson. 1998. Taking It Like a Man: Masculinity in 4 Maccabees. *Journal of Biblical Literature* 117(2): 249–273.

Oxford University Press. 1991. *The New Oxford Annotated Bible* (New Revised Standard Version). Oxford.

Peters, J. 1993. Archaic Milet: Daily Life and Religious Customs from an Archaeozoological Perspective. In *Archaeozoology of the Near East*, ed. H. Buitenhuis and A.T. Clason, pp. 88–96. Universal Book Services / Dr. W. Backhuys, Leiden.

Preston, T.R., V.L. Kossilla, J. Goodwin, S.B. Reed (tech. eds.). 1985. *Better Utilization of Crop Residues and By-Products in Animal Feeding: Research Guidelines 1. State of Knowledge*. FAO Animal Production and Health Paper. Food and Agricultural Organization of the United Nations, Rome.

Redding, Richard. 1991. The Role of the Pig in the Subsistence System of Ancient Egypt: A Parable on the Potential of Faunal Data. In *Animal Use and Culture Change*, ed. P.J. Crabtree and K. Ryan, pp. 21–30. MASCA Research Papers in Science and Archaeology, Supplement to Vol. 8. University of Pennsylvania Museum, Philadelphia.

Roden, C. 1996. *The Book of Jewish Food. An Odyssey from Samarkand to New York*. Alfred A. Knopf, New York.

Shennan, S. 1989. *Archaeological Approaches to Cultural Identity*. Routledge, London.

Stager, L.E. 1995. The Impact of the Sea Peoples in Canaan (1185–1050 BCE). In *The Archaeology of Society in the Holy Land*, ed. T.E. Levy, pp. 332–348. Facts on File, New York.

Thompson, T.L. 1992. *Early History of the Israelite People from the Written and Archaeological Sources*. E.J. Brill, Leiden.

Toplyn, M.R. 1994. *Meat for Mars: Livestock, Limitanei, and Pastoral Provisioning for the Roman Army on the Arabian Frontier (A.D. 284–551)*. Ph.D. dissertation, Harvard University. UMI Dissertation Services, Ann Arbor, MI.

Van Beek, G.W. 1993. Jemmeh, Tell. In *The New Encyclopedia of Archaeological Excavations in the Holy Land* 2, pp. 667–674. The Israel Exploration Society, Jerusalem.

Wapnish, P., and B. Hesse. 1988. Urbanization and the Organization of Animal Production at Tell Jemmeh in the Middle Bronze Age Levant. *Journal of Near Eastern Studies* 47(2): 81–94.

Weinstock, J. 1995. Some Bone Remains from Carthage, 1991 Excavation Season. In *Archaeozoology of the Near East II*, ed. H. Buitenhuis and H.-P. Uerpmann, pp. 113–118. Backhuys Publishers, Leiden.

Wolf, E. 1982. *Europe and the People Without History*. University of California Press, Berkeley.

——— 1990. Facing Power: Old Insights, New Questions. *American Anthropologist* 92(3): 586–596.

PIGS IN ANCIENT EGYPT

Richard A. Lobban

*Department of Anthropology, Program of African and Afro-American Studies,
Rhode Island College, Providence, RI 02908*

Introduction

This chapter seeks to survey the social, mythological, religious, and ecological context of pigs in ancient Egypt. Inevitably it also reflects upon the origins of the prohibition against eating pig flesh which still remains today among Judaic and Islamic populations having origins in the region. Elsewhere I have developed some of the complex integrated and interdisciplinary frameworks which are needed to fully explore this topic (Lobban 1994). This chapter is a revised and up-dated version of that work, abridged so that here I can focus primarily on the presence, role, and circumstance of pigs in ancient Egypt. Other chapters in this present volume, especially that of Hesse and Wapnish, also add to this discussion.

Egypt provides very early evidence of domestic pigs, and it may offer the first recorded inscription recognizing Hebraic populations in the Middle East, at Medinat Habu near Luxor. Thus, it seemed appropriate to investigate Egypt as the potential origin of the Mosaic taboo against swine flesh. In fact, this has been widely accepted for many years (e.g., Murdock 1959:19; Zeuner 1963:262). The historical reconstruction of the Jewish prophet Moses places him in the role of a high-ranking civil servant at some time during the New Kingdom (starting in the sixteenth century B.C.E.), perhaps, more specifically, during the reign of Pharaoh Ramses II (1279–1212 B.C.E.). Given this scenario, Moses would thus have been exposed to contemporary upperclass food preferences and prohibitions. At least by the time of the New Kingdom there was likely an informal taboo on swine flesh for the pharaonic nobility, while at the same time, pork was still widely eaten by poorer people. In the context of conflict within the Egyptian ruling class, Moses fled with his followers into the wilderness of the Sinai, but saw no reason to abandon this already established taboo. In fact, as a marker of "civilized" social status and of Jewish ethnic "borders," the strict observance of the taboo on swine flesh was given supernatural sanctions. It also would be possible to argue, like Zeuner (1963:261), that in comparison with settled agriculturalists who raised pigs, ancient pastoral nomads, such as Jews and Arabs whose economic lifestyles would preclude extensive pig-raising, created a supernaturally backed value system which endorsed their prohibition against pig flesh. As Arab culture came to dominate the entire Middle East, this only became more exaggerated.

Pigs and Ecology in Early Egypt

Pigs probably entered Africa through Egypt from Asia, but it is not clear if they migrated on their own in wild forms, or if they were already in domesticated forms and were driven or carried across the Sinai, or brought over in small coastal boats by their owners. Moreover, domesticated forms may have reverted to feral forms. Houlihan (1996:25) states that Egyptian pigs were domesticated from wild forms already in Egypt. Whatever the case, the environment in the pre-Dynastic Egyptian Delta was ideally suited for the domestication of pigs from wild stock (Flannery 1983:181); perhaps as early as 5000 B.C.E., some domesticated pigs were established in Egypt (Boessneck 1988). Gilman (1976), working in Morocco, has found pig bones at Ashakar dated to the seventh millennium B.C.E., but it is assumed that these were wild animals. Part of this debate revolves around the mandibular and dental indexes used to define domestication and the relative ease with which early pigs reverted to wild conditions. The most convincing proof may exist in data from archaeological assemblages in which there is a relatively high frequency of pig bones, especially those of young pigs which strongly indicate domestication rather than utilization of wild sources. To add to the confusion, some early populations made use of both wild and domesticated forms.

Bones of pigs have been found at pre-Dynastic

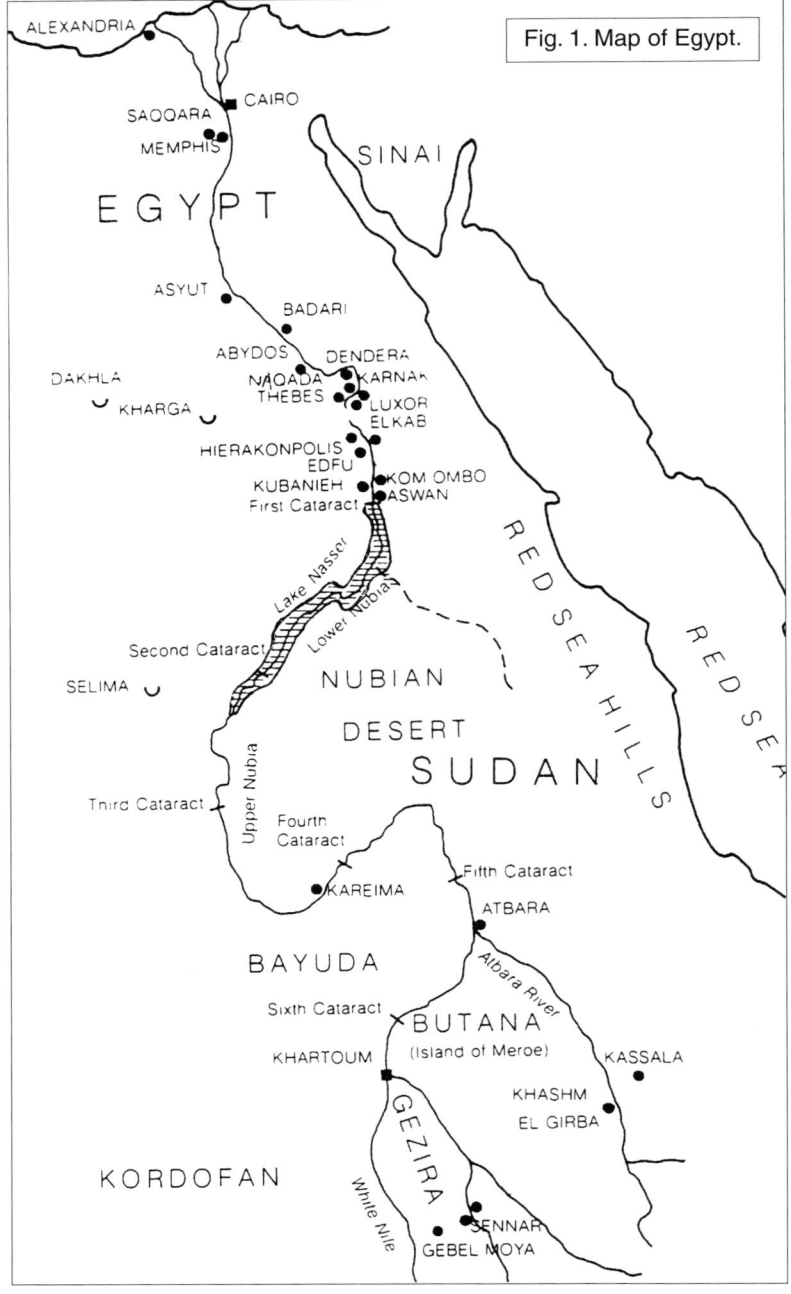

Fig. 1. Map of Egypt.

Toukh in Upper Egypt in about 3400 B.C.E. Hays (1984:68) claims that domesticated pigs were present at El Khattara, some 18 miles north of Luxor in Upper Egypt, in pre-Dynastic times. Houlihan (1996:12) states that domestic pigs were most likely present from the very beginning of the pre-Dynastic Amratian period. Certainly at the close of the Mesolithic, the harvesting and grinding of seed grains in the pre-Dynastic cultural horizons of the Amratians, Badarians, Gerzeans, Nagadans, Tasians, and the Nubian A-Group had taken place (Adams 1976).

So it was in these early Neolithic times in the Sahara that a riverine Nile people with a semi-sedentary lifestyle emerged. Sophisticated hunting and fishing techniques also developed during this period, with a "broad spectrum revolution" in the use of natural foods and materials and a still substantial role for hunting in the adjoining savanna and desert. Prey certainly included pigs, as has just been documented.

Hassan (1984:59–62) agrees that there is the possibility of pig domestication in the Fayum pre-Dynastic era and that they were certainly hunted. However, the relative rarity of pig bones in the Fayum in these early strata can also be interpreted to mean that only wild varieties are represented. Thus, even the experts have a basis to disagree in marginal cases judged by frequency of pig bones in the archaeological record or even in marginal cases judged by anatomical features of the remains. Even when there is agreement about the degree of domestication, there can still be debate over the issue of foreign introduction or local domestication. Thus, for the archaeological data on early pig bones there can be room for interpretation; the final solution to this puzzle is not yet at hand.

Egyptian sites at al-Ma'adi and Merimde. On the northern shores of Lake Qarun in the Fayum Oasis, there is clear proof of a people whose economy was based on barley and emmer wheat, as well as domesticated animals including, claims Kees (1961:32), pigs. At Hemamieh, near Badari in Upper Egypt, in about 4000–3500 B.C.E., sedentary Badarians with pottery and emmer wheat agriculture also had domesticated pigs, sheep, goats, and cattle (Flannery 1961:60–61; 1983:181). The domestic pig (*Sus scrofa* spp.) is also known at the Gerzean site at

In Egypt or elsewhere, domestication is assumed in a context where there is a relatively large percentage of juvenile pig bones, showing a systematic and regular culling. Hunters of wild boars usually have bone assemblages derived from older animals, and a generally lower incidence of pig utilization. At the same time, there were likely reversions from domestic to feral forms where the local habitat was suitable. Flannery (1961:63) states that "Egyptian herd pigs were domesticated out of the native wild stock which abounded in the papyrus marshes of the

Nile..." Murdock (1959:19) states that pigs had certainly arrived in Africa in the early Neolithic and were originally far more important than they are today. Epstein (1971:340–341) has pointed out that even a few pre-Dynastic grave sites in Upper Egypt have clay models of swine. He also notes that faience-glazed sows and boars are found in early strata at Abydos and Hierakonpolis at the start of Dynastic times, when boars appeared on cylinder seals.

Ideally, wild pigs prefer cool and humid conditions. Factors which limit the preferred ecosystem of pigs will reduce their distribution and ability to thrive. They require protection against cold (with fat and hair found on wild boars) and against hot, dry climates (with mud). Pigs have no natural ability to sweat and they have little protective hair, so they often cover themselves with mud to avoid direct sunlight. In the Nile valley and in the adjoining Saharan lands, well before 10,000 years ago and before the Dynastic period began, the climate was clearly much wetter than now. It had permanent sources of water and supported vast numbers of grazing and marsh animals. Hunting and gathering were the only human modes of economic production. Gradually the Sahara began its long process of desiccation; central Saharan rock drawings in the Fezzan and in the Tessili plateau presumably show the beginning of the replacement of hunting and gathering with pastoralism as early as 5,500 B.C.E. and certainly by 3,000 B.C.E. (Vogel 1997:45, 347).

Perhaps as early as 7,000 B.C.E., the desert began to encroach upon the very lush river valley habitat. As the process intensified, the Nile valley became steadily more sharply limited. Carniero (1970) has spoken of the stimulating effect that such impacted habitats can have on domestication and state formation. Wendorf, Schild, and Close (1985:140) propose that the final phase of desiccation of Nubia began in about 3,400 B.C.E. or, in other words, immediately prior to the process of dynastic state formation. Elsewhere I have elaborated on the apparent "triggering" correlation between ecological impaction, state formation, raiding for livestock (especially cattle), and the intensification of agriculture (Lobban 1989).

Livestock Competition in the Nile Valley

In a parallel process of domestication, the ancestors of modern cattle, the aurochs (*Bos primigenius*), appear to have been first domesticated in central Anatolia as early as 6,000 B.C.E. (Perkins 1969). By 3,500 B.C.E., either by diffusion and evolution of this strain or by independent domestication of another long-horned variety, cattle became widespread among North African and Nile valley pastoral peoples (Braidwood and Willey 1962:15; Clutton-Brock 1990). There is also evidence from the so-called Khartoum Mesolithic (really, early Neolithic) that goats were also domesticated by about this same period.

As noted above, well before the Egyptian Old Kingdom (ca. 3,100 B.C.E.) the areas for free grazing were already starting to become limited due to the natural desiccation of the Sahara and to increased human settlement and agricultural land use along the Nile valley. As this progressed, free-range grazing had to be augmented by regular production of animal fodder in the form of Egyptian clover, or *berseem* in modern Arabic. Even in modern times, *berseem* (*Trifolium alexandrinum*) remains the most important single crop by cultivated area (Ikram 1980:175). The production of livestock fodder—especially for cattle, sheep, goats, and donkeys—steadily became, and remains, the major emphasis for Nile valley farmers.

In anthropological studies of the rise of the ancient state, the central role of the domestication of plants and animals is well known. It is also increasingly realized that pastoralism did not emerge spontaneously and unilaterally, but in a synergistic interaction with settled populations (Sadr 1991). However, it is often and perhaps incorrectly assumed that it was increased human population density that directly caused an expansion of agricultural production. Instead the need for more agricultural lands and production has often been for livestock, the greatest consumers in Nile valley antiquity, and to a much lesser degree, for people. When history is viewed with human rather than animal eyes, this point is often overlooked.

The "triggering" link between increased animal populations and increased fodder production in Egypt becomes apparent in considering the results of ancient dynastic raids into Nubia. The raids captured thousands of cattle and other livestock, and much smaller numbers of slaves. As a function of these endlessly repeated and well documented raids, and with the influx of animals from tribute and trade, the Egyptian nobility sent out orders to bring more land into fodder production for their valuable livestock. (The food requirements for slaves were far less, absolutely and relatively.) Then, as today, marshes were drained, swamps filled in, irrigation expanded, and the ecology transformed. One of the earliest animals affected by this ecological transformation was the Egyptian pig, both domestic and wild.

Pre-Dynastic Egyptian pigs initially needed little, if any, fodder. The marshy ecosystems of the undrained Delta and the Fayum Oasis provided an ideal habitat for an animal requiring moist soil for protection from the intense sun. But the combination of (1) desertification, (2) the expanding search for grazing land, (3) the transformation of forest and marsh lands to fodder production, and (4) the increasing importance of cattle, all served to

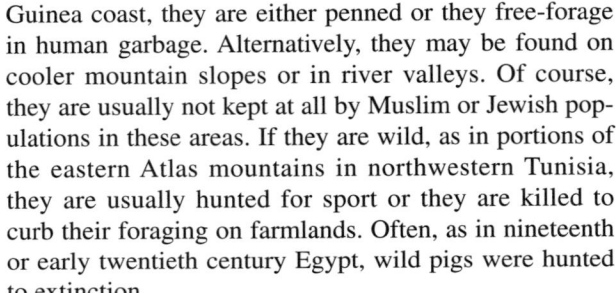

Fig. 2. An image of an Egyptian farmer force-feeding or surrogate-suckling a piglet with milk. After Janssen 1989:34

steadily reduce the pigs' ecosystem. In the expanded search for suitable agricultural ecosystems, humans and their valued cattle competed heavily with pigs.

Initially, the increased production of cattle, sheep, and goats brought ever larger numbers of herdsmen to the formerly less cultivated marshes of the Fayum and the Delta in the summer months. In the winter months, the herders would drive their animals back south along the banks of the Nile, following a regular seasonal pattern of transhumance. Since cattle were held in the highest regard for economic, religious, and cultural reasons, it is ironic that cattle herders were thought of as a rough and lowly pariah group (Erman 1971:438–439); but this is parallel to the status of cowboys and nomads today. Pastoralists in general, especially nomads in arid climates, very rarely herd pigs, even if they are not Jews or Muslims (Harris 1974:34–35). When modern pigs can be found in arid African lands, such as in the Saharan climate of the Cape Verde Islands or among non-Islamic peoples on the Guinea coast, they are either penned or they free-forage in human garbage. Alternatively, they may be found on cooler mountain slopes or in river valleys. Of course, they are usually not kept at all by Muslim or Jewish populations in these areas. If they are wild, as in portions of the eastern Atlas mountains in northwestern Tunisia, they are usually hunted for sport or they are killed to curb their foraging on farmlands. Often, as in nineteenth or early twentieth century Egypt, wild pigs were hunted to extinction.

In this ecological competition, pigs were usually the losers. As long as there were moist marshes their numbers could be sustained at the levels once seen in the wild. However, as they were trapped by the encroaching desert and pushed strongly by the marsh-draining herders and farmers seeking cattle-grazing land, then the future of the free-foraging or even domesticated pigs became limited. They had no advantages over cattle as beasts of burden and traction nor were their hides as useful as cowhides and sheepskins; their milk was not used, and their meat and rare sacrifice to the gods conferred little status.

Despite the early ecological limitations faced by Egyptian pigs, it would be a mistake to conclude that they vanished from the scene. In the XII Dynasty (Middle Kingdom, 2050–1786 B.C.E.), a steward's stela at Abydos describes the people and animals he oversaw, including pigs; typically, however, the pigs were at the bottom of the list (Janssen and Janssen 1989:35). The temple to Osiris at Abydos fairly regularly received pigs as

Fig. 3. An Egyptian swineherd from a tomb at El-Kab in Upper Egypt. After Simoons 1994:15

tribute. And during the time of Sesostris I, the nomarch Hepdjefi at Assyut received taxes in the form of sacrifices of low-status goats and pigs (Kees 1961: 91).

During the New Kingdom (1568–1080 B.C.E.), the domestic pig is portrayed at Thebes and other places (Fig. 2). At this time, emmer wheat was reported as pig food (Kees 1961:88). In one text from the XVII Dynasty, the nomarch Renni of El Kab, in Upper Egypt, reported that he owned, or received as tribute, 122 cattle, 100 sheep, 1200 goats, and 1500 pigs (Kees 1961:87). In the XVIII Dynasty, a stela records that the chief steward of Amenophis III offered 1,000 pigs to the lower temple personnel of his master; at Tel al Amarna, excavations have revealed farrowing pens for pigs; and in most towns of the New Kingdom there are plenty of pig bone remains (Janssen and Janssen 1989:34). Nonetheless, pork was "poor man's food." Drawings in the tomb of Paheri, also in Kab, depict a swineherd at work (Fig. 3). Even in the late Saite or XXVI Dynasty in Egypt, sows were depicted on faience amulets (Fig. 4). In the glorious XVIII Dynasty (1570–1305 B.C.E.), swine were depicted in various scenes including their use in trampling seeds at planting time.

Fig. 4. A pig amulet perhaps to ward off evil.
After Epstein 1971:341

The Greek writers Herodotus and Eudoxus also reported this Egyptian practice of sowing seeds and then deliberately herding pigs onto the field to trample the seeds into the fertile soil. During Greco-Roman times in Egypt (333 B.C.E.–A.D. 324), the taboo against swine flesh was generally not observed and evidently large numbers of pigs were raised for food. In the case of contemporary Meroë in adjoining Nubia, Shinnie and Bradley (1980:310) report from their archaeological work the almost complete absence of bones of domestic pigs; for those very few *Suidae* bones present, a case can be made for their being wild wart hogs and not domestic pigs at all. Since the people of Meroë generally followed ancient Egyptian religious traditions, one might assume that the prohibition against pigs may still have been in force there. Pre-Islamic Berbers, some Rifian Moroccans, Neolithic Guanche in the Canaries, as well as North African European populations (other than Jews), all raised and ate pigs (Epstein 1971:330–331). The arrival of Islam across North Africa in the seventh and eighth centuries A.D. probably reduced pig populations considerably, yet Islam traveled much more slowly southward across the desert and up the Nile. In Nubia, the medieval Christians raised considerable numbers of pigs, having no prohibition against them. During the Arab invasion of Ibrim in Nubia in 1173 by Turan Shah (the brother of Saladin el Ayyubi), some 700 pigs were reported slain. Pigs are still raised in the Nuba Mountains of the Sudan today.

The pre-Nilotic ancestors of the Sudanese Funj Sultanates raised swine and may have, in fact, developed a local pig breed (*Sus scrofa sennarensis*). Greek and Coptic villages and merchants in the Sudan continued to raise pigs for markets during Islamic times and down to the present. James Bruce, the Scottish explorer of the Sudan in the late eighteenth century, reported ironically that "hog's flesh is not sold in the market; but all the people of Sennar eat it publicly; men in office, who pretend to be Mahometans [sic], eat theirs in secret" (1798:371).

Wild boars were still reported in Lower and Middle Egypt, in Delta marshes, and in the Fayum Oasis until the end of the nineteenth century. The last known case of a wild pig shot was in 1902, and throughout the nineteenth century there were reports of crop damage by pigs. In 1846, for example, a substantial campaign was launched by 832 soldiers and 19 officers which resulted in the killing of 756 wild pigs in Gharbiya and Menufiya and another 104 in Sharqiya and Dakhliya (Epstein 1971:326). Wild boars (*sangliers*) are still hunted today in Tunisia; they are sometimes eaten after the hunt.

At present, Sudanese and Egyptian Copts raise pigs for their own consumption and for sale. Pigs are still kept in Cairo by the *zabbaleen* (garbage collectors). The *zabbaleen* originate in the Christian communities of the Assyiut governorate in Middle Egypt. They moved to areas in greater Cairo in the 1940s, but by the mid-1960s local officials opposed their practice of raising pigs and forced them to move to more remote locations, such as the vast burning dumps near the Muqattam Hills where they still reside (Oldham, El-Hadidi, and Tamaa 1987).

Without the severe ecological restrictions of the Lower Nile, many non-Islamic populations of the Upper Nile in the southern Sudan and the Nuba Mountains also raise pigs today. More often they are raised by the technique of free-foraging rather than herding. Usually they are consumed domestically or locally rather than sold to a market. Thus, despite the Islamic taboo, pigs are found and raised in numerous places on the African continent, especially in the non-Islamic regions in coastal west, central, and south Africa. The minority ethnic position of pig herders, the peripheral nature of the pig economy, and the low status of those who trade in pigs follow trends established thousands of years earlier (Hecker 1982; Redding 1991).

Pigs in Ancient Egyptian Mythology

Long before the appearance of the Judaic or Islamic taboo, one can see a long history of negative images and roles for pigs in Egypt. Parallel to the competition between cattle and pig for a suitable ecosystem in the Delta was the conquest by the falcon-totem nomarchs of Upper Egypt of the pig-totem nomarchs in the Lower Egyptian Delta. The pig, represented by the god Seth, was forever devalued in this historic contest. One can imagine, says Kees (1961:37), that the falcon-totem nomarchs would forbid sacrifice of pigs as inappropriate to their god, Horus.

Although pigs were raised and eaten in pre-Dynastic times, there is no evidence that they were shown any special reverence. Smith (1969:311) says that the same was true in Dynastic times, although he does claim that they were sacrificed, but as low-status animals. A measure of this low status was expressed by Herodotus who wrote that swineherds were not allowed to participate in the important Thinite Sed festivals.

The main mythological link to the low status of pigs rests with the god Horus' opposition to swine because of their close association with the god Seth (or variously Set, Suth, and Setesh). In Egyptian mythology, Seth, one of the four children of Geb and Nut, had treacherously murdered his older brother, Osiris, who was the father of Horus. Horus was the patron god of all pharaohs. Seth, who has been portrayed in the form of a pig, also blinded one of Horus' eyes. In the wake of these crimes, Horus assumed the eternal responsibility of seeking revenge for his father's death. Presumably this moral tale taught the lesson of the relentless pursuit of the forces of good over the forces of evil. Just as Seth's brother Osiris married his sister Isis, Seth married his other sister, Nephthys. This primordial creation myth of good versus evil has an important parallel with the fratricide of Cain and Abel in Christian theology, which finds many roots in the Nile valley. The appearance of evil Seth could be represented by the crocodile, *at*; the hippopotamus, *khab*; or the pig, *rer* (Fig. 5). It appears that the sneaky sexual and bi-sexual nature of Seth is even associated with the complexities of pharaonic

Fig. 5. The Egyptian determinative, transliteration, and hieroglyphs for "pig."
After Budge 1920, 2:428a

power in the case of the Egyptian '*w3s*' scepter, which represented the staff of Thebes and was commonly carried as a symbol of important deities (Lobban and Sprague 1997).

Upon judgment day (another ancient Egyptian concept), the *ba* (soul) would be appraised by Osiris and other impaneled gods and if one's life was judged poorly, one would be thrown into the awaiting jaws of *at* or *khab*; this fate was made even more fearsome by the potential for spending eternity in their stomachs. Pigs seldom had this exact mythological function. However New Kingdom and Late Period tomb drawings sometimes show a pig being carried away on a boat on judgment day (Fig. 6); the assumption is that the soul being judged was not wicked, so the pig could absorb any evil forces and the soul could pass happily to the 'other world.'

In any case, the role of pigs was somewhat more benign than that of hippopotami. The hellish role cast for *at* and *khab* may help to justify or explain their extinction in Egypt. All three were water- or marsh-loving animals whose enemies were the desert sun and drying winds, which were likewise represented by the complex natures attributed to Seth (Thomas 1986:60).

Probably as early as the Old Kingdom, Seth was associated not only with pigs, but also with connotations of evil and deception. In the Book of the Dead, Erman (1971:441) notes that Seth actually assumes the form of a pig, one of his sacred totems. In tombs, the deceased is occasionally shown spearing Seth the pig (Fig. 7); this is to ensure that the deceased's soul will not be carried away by the force of evil (Epstein 1971:342). Seth in the form of a hippopotamus is also seen being harpooned in the head and testicles by Horus (Budge 1987:361).

Seth was worshiped especially from the V to XIX Dynasties. He represented and controlled the destructive power of the sun's heat, which is the enemy of the water-loving mammals he also symbolized. This contradiction of protection and destruction is typical of his duplicitous and treacherous nature. As if to be especially provocative, the Asiatic

Fig. 6. Two baboons representing the god Thoth driving a pig away from a judgment day scene in which the pig symbolized the evil god Seth.
After Simoons 1994:19

Hyksos invaders of Egypt (Dynasties XV and XVI in the seventeenth century B.C.E.) adopted Seth as one of their favored gods. To Egyptians, this insult offered further proof of Seth's evil and treacherous nature. In the subsequent New Kingdom, when the glorious Egyptian empire was rebuilt, the links between evil Seth and the foreign conquerors were not forgotten.

Sometime between the XXI and XXV Dynasties (1080–664 B.C.E.), in the wake of the great Rameside epoch and the contemporary departure or flight of the Jews led by Moses, a violent reaction took place against Seth worship. Identified as the embodiment of evil and the opponent of good, statues and figures of Seth were

Fig. 7. At judgment day an attendant spears an evil pig. After Epstein 1971:342

smashed and his image was defaced. This extreme reaction may also explain the paucity of pig representations to a certain degree. Virtually no bronze or faience figures of Seth are found after this period, although pig representations in faience may be an exception.

Budge (1926:116) noted that pigs were kept in "tolerably common" numbers in ancient Egypt. Both pigs and asses were eaten by slaves and swamp-dwellers. Pork was eaten in the state's workers' village in the Theban necropolis during Rameside times, but a prohibition on their flesh was rigorously observed by kings and nobility. Smith (1969:312) adds that pigs were not even hunted by royalty and were not specially kept for fattening, nor offered as special temple tribute or sacrifices. Equally there is no indication that pigs were kept as pets. Subsequently, in classical Greco-Roman mythology, Seth and pigs in general began to be associated with the monster god Typhon. Zeus for the Greeks, or Jupiter for the Romans, waged continual war against the evils of Typhon. Smith (1969:313) considers that these limited or negative roles for pigs prove that their "Typhonian" nature was established, and pigs were not utilized for temple sacrifice because their flesh was already taboo. Other "Typhonic" food prohibitions were placed against the flesh of horses and asses. Indeed, pigs were rather rarely depicted and their bones never found in tombs, and only on very special occasions were they featured in temple offerings.

On the other hand, Herodotus and Plutarch, writing in post-Dynastic times, report that pigs were sacrificed in Egypt on special annual occasions, such as the Sokaris festival at Memphis when pigs were sometimes offered at full moon to Osiris or Seth/Typhon. Perhaps because pigs signified her brother Seth, it was considered by some that pigs were also sacred in special contexts to Isis herself (Sauer 1952:343).

Within Egyptian folklore that has survived until the twentieth century, there is a persistent belief in the ambiguous, yet evil, nature of pigs, as well as the belief that they should be raised with horses for the health of the horses. According to oral history, and occasional modern practice, the presence of pigs helps to ward off, or absorb, evil forces which may otherwise afflict the horses. Pork flesh was also reputed to have medicinal value for horses and asses.

Pigs, Mythology, and Diseases

A rich and diverse literature exists around the symbolism of pig sacrifice and the interpretation of mythology. Readers may be interested in the works of Frazer (1959), Gimbutas (1982), and Walker (1983, 1988). A common explanation for the pig taboo rests in their association with disease; but, whether in mythological cases or occurrences of actual diseases, this explanation must be rejected. Assuming that Smith (1969) is correct that the "Typhonian" nature of pigs was accepted broadly by the late New Kingdom, it is not surprising that the Ptolemaic Greeks who were the next to admire and rule over Egypt would have reached a similar judgment about swine. Within Greek mythology, Typhon, son of Typhoeus, was a monster capable of producing a fever or vapor from which we have inherited the term *typhoid* fever. Actually typhoid fever is caused by a bacillus usually transmitted in foods which cause catarrh, an extreme intestinal infection. Another early notion, propounded by Tacitus, was that leprosy was caused by contact with pigs.

In Book II of his famous history of the Middle East region, Herodotus said that

> The pig is regarded among them as an unclean animal, so much so that if a man in passing accidentally touches a pig, he instantly hurries to the river, and plunges in with all his clothes on. Hence, too, the swineherds, notwithstanding that they are of pure Egyptian blood, are forbidden to enter into any of the temples, which are open to all other Egyptians; and further, no one will give his daughter in marriage to a swineherd, or take a wife from

among them, so that the swineherds are forced to intermarry among themselves. They do not offer swine in sacrifice to any of their gods, excepting Bacchus and the Moon, whom they honor in this way at the same time, sacrificing pigs to both of them at the same full moon, and afterward eating of the flesh...To Bacchus, on the eve of his feast, every Egyptian sacrifices a hog before the door of his house, which is then given back to the swineherd by whom it was furnished, and by him carried away. (Herodotus, Book II, 1956, p. 98)

Others have offered various interpretations about the famed remarks of Herodotus (Epstein 1971; Janssen and Janssen 1989).

The Pork Taboo in the Torah, Bible, and Koran

The formal pork taboo appears in the Old Testament (Torah) probably written in the mid-fifth century B.C.E. In Leviticus 11 (4–47) numerous animals are specified as "clean" or "unclean." These latter include "the swine, though he divided the hoof, and be clovenfooted, yet he cheweth not the cud; he is unclean to you. Of their flesh shall ye not eat, and their carcass shall ye not touch; they are unclean to you."

This clearly establishes the religious prohibition for Jews, and briefly for Christians until the New Testament reforms noted in Acts 10 (9–16) allowed early Hellenized Christians to abandon this food restriction. Later, Muslims, wanting or needing to distinguish themselves from Christians, simply renewed the former taboo against pig flesh.

For Muslims, reference is made to Surah V, the Table Spread, verse 3, "Forbidden unto you (for food) are carrion and blood and swineflesh..." The taboo is repeated in Surah VI, Cattle, verse 145, which indicates that swine flesh "verily is foul or the abomination which was immolated to the name of other than Allah. But whose is compelled (thereto), neither craving nor transgressing, (for Him) lo! your Lord is Forgiving, Merciful." One may conclude that Muslims may even raise pigs, but just not eat them unless "compelled" by emergency circumstances. Islamic practices also exclude scavenging and bottom-feeding animals from the diet.

It is critical for Muslims that the solemn slaughter (*zaba'a*) of animals be done with an invocation to Allah and that the blood flows, that is, the animal should not have died beforehand because of being beaten, strangled, or gored, or have met death in circumstances which would cause it to be carrion. Animals sacrificed to gods other than Allah are not to be eaten. In the case of hunting with falcons, Muslims ought to eat the game pursued and the name of Allah should be invoked when the falcon is released (Ali 1977:240–241, 330–333).

Discussion

Certainly we must conclude that any strictly monocausal explanation of the rise and evolution of the pig taboo—one based solely on religion, ecology, ancient history, disease theory, or mythology—will not be sufficient to understand this practice. However, the main thesis here is that all of these factors together do provide the basis for a new, integrated perspective on this very old problem. I cannot agree with the statements of Clutton-Brock that "nobody knows why one species of animal is favoured as a supplier of meat to one nation, whilst to another it is considered a taboo animal, unclean and untouchable" and "There is no satisfactory explanation at present as to why they were later so abhorred" (1981:76–77).

From the evidence provided above, several explanations for the pig taboo have been presented. This paper adopts a framework incorporating various perspectives after concluding that the idealist explanations offered by symbolic/supernatural dichotomies such as those of Douglas (1966), or the ethnoscientific approach by Hunn (1979) typically turn into functionalist tautologies which may explain the practice at present, but do not shed much light on its origins. This conclusion also rejects the despairing viewpoint of Clutton-Brock which determines that this problem will never be solved.

Instead, by turning toward livestock competition, and historical, religious, mythological, political, and ecological data such as explored by Diener and Robkin 1978; Harris 1974, 1977, 1985; Rappaport 1967; Shnirelman 1988; and Vayda 1968, 1971, we have made more concrete progress in resolving this issue. In these works, the discussion has turned toward the complex historical evolution of the taboo. Despite important disagreements of interpretation and emphasis, all of these latter scholars would likely agree that there is a relationship between pig demography within its regional ecosystem and the taboo. There is a need to control relative population size of pigs by sacrifice, taboo, hunting, or other social, religious, or ideological restraints. Without some restraints, the need to produce food for pigs would finally exceed human capacity or interest in meeting this need.

The common solution to the problem of controlling the domestic pig population is to eat the vast majority of piglets and young pigs before they are one year of age. Care is taken to preserve only a favored boar and a number of sows for farrowing. This has been the approach of most pig producers from ancient times until the present (DuCos 1969:271), especially in moist, temperate areas which have no other special restrictions on pig production. The Egyptian or Middle Eastern case became confounded by environmental peculiarities, a prodigiously convoluted mythology, and a complex ethnic and religious context, all over an extremely long period of time in an area which became seminal in forming three great world religions.

In a review of the status of pigs from a world perspective we find that in Melanesian and New Guinea societies that lack other major livestock, pigs became highly valued for social status. These cultures also faced the problem of valuing pigs, but needing to limit their population. There can be little doubt that the wars to either seize or kill pigs, and the huge pig feasts all rest upon various configurations of pig demography, as Harris (1974:42–49) would likely agree.

Waddell (1972:204, 208–209) states that there are abundant examples of peoples in Highland New Guinea who maintain large, dependent pig populations supported by extensive taro gardens. He notes that "While intensification is in part a response to an expanding human population, an essential concomitant is undoubtedly the large, dependent pig population" (1972:210). Sillitoe, also writing on New Guinea, says that the amount of produce which the Wola feed to their pigs is surpassing, especially in comparison to per capita human consumption (1983:228). In his detailed survey of consumption of crops he found that pig consumption was about three times that of adult humans (1983:229). In studying the Etoro, Kelly (1988) discovered a major involvement in raising food, especially sago, for feeding domestic pigs. This is done, in part, for the ritual and social importance of exchanging pigs and pork. The higher the density of human populations, the greater the effort which must be made to produce food for the pigs. It is only in lower-density populations that free-foraging is possible, according to Kelly.

Rappaport (1967:59) confirms that large pig herds require large-scale cultivation to produce harvests "especially for pigs, that is, to work for the pigs and perhaps to give them food fit for human consumption." Indeed, his study showed that pigs received 53.7% of the sweet potato crop and 82.0% of manioc production (1967:60). Thus, there are high costs in pig maintenance, especially in sweet potato and manioc farming. Pigs are not only fed a surplus from gardens for human needs and other vegetal surpluses and waste, but increased acreage may be needed just for production of pig fodder (Rappaport 1967:65). Finally, Watson (1977) has also shown that the widespread introduction of sweet potatoes in New Guinea has allowed for a rapid expansion of the previously smaller numbers of domestic pigs. Now great effort and resources are expended to produce sweet potatoes for fodder to expand the pigherd size for social prestige and food needs.

In the case of China, where pigs are very popular as food and no taboo exists, hundreds of thousands are slaughtered annually, and in the New Guinea case cited by Rappaport (1967), where pigs actually confer high status, there are cyclical slaughters of hundreds on a single day. Without such measures, the human population would soon be spending a very large proportion of its time just feeding and supporting its pigs.

The history of domesticated pigs in Egypt clearly places these animals in the moist soils, river banks, and marshes of Egypt well before Dynastic times. With a relatively sparse population of hunters and fishermen, the pigs thrived as a domestic and wild species; but as agriculture and irrigation were intensified for increased production of cattle, sheep, and goats, the pressure mounted to drain and otherwise incorporate the ideal habitat of the pigs and transform it to other types of cultivation.

In these early times, pigs were rather numerous and available as a popular food. It was common for certain regions of Egypt to be known by the animals which occupied them, thus it was not surprising that the Delta and Lower Egypt had a pig totem. However, since the impulse for ancient state formation probably came from Upper Egypt which had its own totems, especially the royal falcon Horus, the pig was socially and politically devalued. Indeed, it was probably early in Dynastic times that pigs were mythologically and politically linked to Seth, who, in this form or others, became the embodiment of evil and treachery.

Thus, as the major ecological transformation unfolded, perhaps in association with a political or even military defeat for the Delta nomarchs represented by the pig totem, then swine began to be less and less desirable by comparison with other livestock. The early growth of Egyptian religion and mythology anchored and embellished these perceptions. As Egyptian society passed through its millennia, the status of pigs slipped further, from indifference to avoidance and then to prohibition and taboo. This trajectory was aided when the foreign "shepherd king" Hyksos invaders of Egypt adopted Seth as one of their gods. This act thereby added to the future repugnance of pigs when the Egyptian people were later able to restore their empire to its greatness.

When the Jewish leader Moses came onto the scene, he accepted the dietary values of the Egyptian ruling class he served. Once these "revelations" were written, it subsequently led to the continued observance of the taboo for the Jewish people. One might wonder if Jewish or Muslim capitals had been located in cooler, moister, or more temperate climates for a millennium or so, would the observance of the "revealed" taboo against swine also slip? If we look at modern France for some data, we find that far more pigs are raised in the cooler north than in the warmer and drier south (Duckham and Masefield 1970:268–269). In modern tropical sub-Saharan Africa, great numbers of pigs are raised even in areas under Islamic influence. It may also be seen that some of the European or Ashkenazi Jews are somewhat more liberal relative to food taboos than are Sephardic Jews living in the Middle East. In the case of Christians, one may also wonder if the relocation of the capitals of Christendom to cooler Rome, rather than the desert environment of Mecca or Cairo, may have been a factor in the loss of the early Christian observance of the taboo.

To conclude, the process of ancient Egyptian state formation included processes of ecological transformation, livestock competition, mythological evolution, and territorial conquest. Pigs, a totemic symbol of the conquered Delta, were defeated by the falcon-god Horus, the totem of Upper Egypt and of all subsequent pharaohs. There was the mythological rivalry between pigs, in the form of Seth, and Horus who was endlessly pursuing Seth, the murderer of his father, Osiris. This is one of the very oldest and enduring themes of ancient Egyptian religion which continues in the eternal combat of good and evil in Judaism, Christianity, and Islam. Through the centuries Seth became increasingly identified with all elements of treachery and evil. A late step in this progression was the Hyksos' acceptance of Seth, which for the Egyptians added still another reason to find Seth loathsome.

At least for the rulers and their allies, both Seth and pigs were thoroughly disgusting. Meanwhile, out of need or preference, pigs were still raised, donated, consumed, and extensively found throughout the Nile valley. It was in these centuries, around the time of Ramses II or III, that Hebraic populations coalesced around a clear ethnic identity, and perhaps because of their link to, or admiration of, Egyptian aristocracy they also accepted and codified the taboo. The subsequent transmission of the taboo went on from the Egyptians to the Jews and was observed by the early Christians until some religious reforms were introduced which allowed the consumption of pig flesh. With the birth of Islam in the seventh century, the taboo was reborn in a simplified and refortified version which has become emblematic of this new religion until the present.

In general, taboos exist to control access to something otherwise desirable, such as specific sexual partners or foods. This is often achieved by establishing specific metaphysical parameters between the domains of sacred and profane as the idealists have noted. In the case of potentially edible, delicious pigs in Egypt, the changing ecology and economy built around drainage of swamps for increased fodder production and cattle-grazing meant a loss for pigs, and importantly, a genuine and increasing need to control pig populations. This is just one of several alternative strategies adopted by humans. These include: (1) regularly eating or sacrificing the largest portion of suckling or young pigs, (2) destroying pig ecosystems and habitat, (3) actively hunting wild or feral Suidae until extinction or low populations are achieved, and (4) excluding or marginalizing them as a socially desirable food through the creation of socio-religious taboos which would severely restrict pig production and consumption. Egyptians used a combination of the last three strategies.

Acknowledgments
Appreciation is gratefully acknowledged for practical assistance, constructive comments, and reflective criticism of early forms and parts of these ideas. Especially helpful have been Carolyn Fluehr-Lobban, Terence E. Hays, Andrew Rowan, Calvin Schwabe, and Victor Shnirelman.

References
Adams, William Y. 1976. *Nubia: Corridor to Africa.* Princeton University Press, Princeton.

Ali, A. Yusuf. 1977. *The Glorious Koran: Translation and Commentary.* American Trust Publications, Washington, DC.

Boessneck, J. 1988. *Die Tierwelt des Alten Aegypten.* C.H. Beck, Munich.

Braidwood, Robert J., and Gordon R. Willey (eds.). 1962. *Courses Toward Urban Life.* Aldine, Chicago.

Bruce, James. 1798. *An Interesting Narrative of the Travels of James Bruce into Abyssinia to Discover the Source of the Nile.* S. Etheridge, Boston. (American abridged version.)

Budge, E.A. Wallis. 1920. *An Egyptian Hieroglyphic Dictionary.* Reprint, 2 vols. Dover, New York, 1978.

——— 1926. *The Dwellers on the Nile.* Reprint. Dover, New York, 1977.

——— 1987. *The Mummy: A Handbook of Egyptian Funerary Archaeology.* 2d ed. Reprint. Routledge and Kegan Paul, New York.

Clark, J. Desmond, and Steven A. Brandt (eds.). 1984. *From Hunters to Farmers.* University of California Press, Berkeley.

Clutton-Brock, Juliet. 1981. *Domesticated Animals from Early Times.* University of Texas Press, Austin.

——— 1990. *The Walking Larder.* Heinemann, London.

Diener, P., and E.E. Robkin. 1978. Ecology, Evolution, and the Search for Cultural Origins: The Question of Islamic Pig Prohibition [with discussion and commentary]. *Current Anthropology* 19(3): 493–540.

Douglas, Mary. 1966. *Purity and Danger: An Analysis of the Concepts of Pollution and Taboo*. Routledge and Kegan Paul, London.

Duckham, A.N., and G.B. Masefield. 1970. *Farming Systems of the World*. Praeger, New York.

Ducos, P. 1969. Methodology and Results of the Study of the Earliest Domesticated Animals in the Near East (Palestine). In *The Domestication and Exploitation of Plants and Animals*, ed. P. Ucko and G.W. Dimbleby, pp. 265–275. Aldine, Chicago.

Epstein, H. 1971. *The Origin of the Domestic Animals of Africa*. 2 vols. Rev. ed. Africana Publishing, New York.

Erman, Adolf. 1971. *Life in Ancient Egypt*, trans. H.M. Tirard. 2d ed. Dover, New York.

Flannery, Kent V. 1961. *Skeletal and Radiocarbon Evidence for the Origins of Pig Domestication*. M.A. thesis, Department of Anthropology, University of Chicago.

——— 1983. Early Pig Domestication in the Fertile Crescent: A Retrospective Look. In *The Hilly Flanks and Beyond. Studies in Ancient Oriental Civilization*, ed. T.C. Young, Jr., P.E.L. Smith, and P. Mortensen, pp. 163–188. Oriental Institute, Chicago.

Frazer, James George. 1959. *The New Golden Bough*. Criterion Books, New York.

Gilman, A. 1976. *A Later Prehistory of Tangier, Morocco*. American School of Prehistoric Research, Bulletin 29, Peabody Museum, Cambridge, MA.

Gimbutas, Marija. 1982. *The Goddesses and Gods of Old Europe: Myths and Cult Images*. University of California Press, Berkeley.

Harris, Marvin. 1974. *Cows, Pigs, Wars, and Witches: The Riddles of Culture*. Vintage Books, New York.

——— 1977. *Cannibals and Kings: The Origins of Cultures*. Random House, New York.

——— 1985. *Good to Eat: Riddles of Food and Culture*. Simon and Schuster, New York.

Hassan, Fekri A. 1984. Environment and Subsistence in Predynastic Egypt. In *From Hunters to Farmers*, ed. J. Desmond Clark and Steven A. Brandt, pp. 57–64. University of California Press, Berkeley.

Hays, T.R. 1984. A Reappraisal of the Egyptian Predynastic. In *From Hunters to Farmers*, ed. J. Desmond Clark and Steven A. Brandt, pp. 65–73. University of California Press, Berkeley.

Hecker, H.M. 1982. A Zooarchaeological Inquiry into Pork Consumption in Egypt from Prehistoric to New Kingdom Times. *Journal of the American Research Center in Egypt* 19:59–69.

Herodotus. 1956. *The History of Herodotus, Book II*, ed. Manuel Komroff. Tudor Publications, New York.

Houlihan, Patrick F. 1996. *The Animal World of the Pharaohs*. Thames and Hudson, London.

Hunn, Eugene. 1979. The Abominations of Leviticus Revisited. In *Classifications in Their Social Context*, ed. Roy F. Ellen and David Reason, pp. 103–116. Academic Press, New York.

Ikram, Khalid. 1980. *Egypt: Economic Management in a Period of Transition*. Johns Hopkins University Press, Baltimore.

Janssen, Rosalind, and Jack Janssen. 1989. *Egyptian Household Animals*. Shire Egyptology 12. Princes Risborough, Aylesbury, Bucks, UK.

Kees, Hermann. 1961. *Ancient Egypt: A Cultural Topography*. The University of Chicago Press, Chicago.

Kelly, Raymond. 1988. Etoro Suidology: A Reassessment of the Pig's Role in the Prehistory and Comparative Ethnology of New Guinea. In *Mountain Papuans: Historical and Comparative Perspectives from New Guinea-Fringe Highland Societies*, ed. James F. Weiner, pp. 111–186. University of Michigan Press, Ann Arbor.

Lobban, Richard A., Jr. 1989. Cattle and the Rise of the Egyptian State. *Anthrozoös* 2(3): 194–201.

——— 1994. Pigs and Their Prohibition. *International Journal of Middle East Studies* 26(1): 57–75.

Lobban, Richard A., Jr., and Michael Sprague. 1997. Bulls, and the 'w3s' Sceptre in Ancient Egypt and Sudan. *Anthrozoös* 10(1): 14–22.

Murdock, George Peter. 1959. *Africa: Its Peoples and Their Culture History*. McGraw Hill, New York.

Oldham, Linda, H. El-Hadidi, and H. Tamaa. 1987. *Informal Communities in Cairo: The Basis of a Typology*. Cairo Papers in Social Science Vol. 10, no. 4. American University in Cairo Press.

Perkins, Dexter, Jr. 1969. Fauna of Çatal Hüyük: Evidence for Early Domestication in Anatolia. *Science* 164 (3876): 177–179.

Rappaport, Roy A. 1967. *Pigs for the Ancestors: Ritual in the Ecology of a New Guinea People*. Yale University Press, New Haven.

Redding, Richard A. 1991. The Role of the Pig in the Subsistence System of Ancient Egypt: A Parable on the Potential of Faunal Data. In *Animal Use and Culture Change*, ed. P.J. Crabtree and K. Ryan, pp. 21–30. MASCA Research Papers in Science and Archaeology, Supplement to Vol. 8, University of Pennsylvania Museum, Philadelphia.

Sadr, Karim. 1991. *The Development of Nomadism in Ancient Northeast Africa*. University of Pennsylvanian Press, Philadelphia.

Sauer, C.O. 1952. *Agricultural Origins and Dispersals*. American Geographical Society, New York.

Shinnie, Peter L., and Rebecca J. Bradley. 1980. *The Capital of Kush 1: Meroë Excavations 1965–1972*.

Meroitica 4. Akademie-Verlag, Berlin.

Shnirelman, Victor A. 1988. Primitive Warfare and Ideology: Why the Maring Avoid Seizing Pigs. *Human Peace* 6(3): 3–7.

Sillitoe, Paul. 1983. *Roots of the Earth: Crops in the Highlands of Papua New Guinea.* Manchester University Press, Manchester.

Simoons, Frederick J. 1994. *Eat Not This Flesh: Food Avoidances from Prehistory to the Present.* 2d ed., rev. and enl. University of Wisconsin Press, Madison.

Smith, H.S. 1969. Animal Domestication and Animal Cults in Dynastic Egypt. In *The Domestication and Exploitation of Plants and Animals,* ed. P. Ucko and G.W. Dimbleby, pp. 307–314. Aldine, Chicago.

Thomas, Angela P. 1986. *Egyptian Gods and Myths.* Shire Egyptology 2. Princes Risborough, Aylesbury, Bucks, UK.

Ucko, Peter, and G.W. Dimbleby (eds.). 1969. *The Domestication and Exploitation of Plants and Animals.* Aldine, Chicago.

Vayda, Andrew P. 1968. Hypotheses about Functions of War. In *War: The Anthropology of Armed Conflict and Aggression,* ed. M. Fried, M. Harris, and R. Murphy, pp. 85–91. Natural History Press, Garden City, NY.

——— 1971. Phases of the Process of War and Peace Among the Maring of New Guinea. *Oceania* 42(1): 1–24.

Vogel, Joseph O. 1997. *Encyclopedia of Precolonial Africa.* AltaMira Press, Walnut Creek, CA.

Waddell, Eric. 1972. *The Mound Builders: Agricultural Practices, Environment, and Society in the Central Highlands of New Guinea.* University of Washington Press, Seattle.

Walker, Barbara G. 1983. *The Woman's Encyclopedia of Myths and Secrets.* Harper and Row, New York.

——— 1988. *The Woman's Dictionary of Symbols and Sacred Objects.* Harper and Row, New York.

Watson, James B. 1977. Pigs, Fodder and the Jones Effect in Postipomoean New Guinea. *Ethnology* 16(1): 57–70.

Wendorf, Fred, Romuald Schild, and Angela E. Close. 1985. Prehistoric Settlements in the Nubian Desert. *American Scientist* 73:132–141.

Zeuner, F.E. 1963. *A History of Domesticated Animals.* Harper and Row, New York.

MASCA Research Papers in Science and Archaeology

Series Editor
Kathleen Ryan

Production Editors
Jennifer Quick
Helen Schenck

Advisory Committee
Stuart Fleming, Chairman
Philip Chase
Patrick McGovern
Henry Michael
Naomi F. Miller
Katherine M. Moore
Vincent Pigott

Design and Layout
Jennifer Quick

**Customer Service/
Subscription Manager**
Jennifer Bornstein

Volume 15, 1998

The subscription price for *MASCA Research Papers in Science and Archaeology* is $25, payable in US dollars. We also accept VISA and MASTERCARD. This price covers one main volume per year. In addition, we publish supplementary volumes which are offered to subscribers at a discounted price.

This is a refereed series. All materials for publication should be sent to The Editor, *MASCA Research Papers in Science and Archaeology*. Subscription correspondence should be addressed to The Subscription Manager, MASCA, University of Pennsylvania Museum, 33rd and Spruce Streets, Philadelphia, PA 19104-6324.